6

Student Book

Pauline Rogers
Aaron Tait

Australia • Brazil • Japan • Korea • Mexico • Singapore • Spain • United Kingdom • United States

Contents

Identifying Place

DATE:

1 State the value of the **6** in each number.

a 31 436 ____________ **b** 16 053 ____________

c 140 161 ____________ **d** 243 628 ____________

e 43 061 ____________ **f** 156 112 ____________

2 State the value of the **2** in each number.

a 231.04 ____________ **b** 612.53 ____________

c 320.11 ____________ **d** 243.68 ____________

e 427.60 ____________ **f** 652.163 ____________

3 What is the value of the underlined digit?

a $2\underline{9}016$ ____________ **b** $42\underline{1}19$ ____________

c $168\underline{3}2$ ____________ **d** $6\underline{0}.479$ ____________

e $\underline{1}8.560$ ____________ **f** $\underline{4}006.98$ ____________

4 Write the following numbers in numerals.

a one thousand, nine hundred and fifty-four ____________

b twelve thousand, one hundred and twelve ____________

c sixty-one thousand, four hundred and fifty-one ____________

d one hundred and eight ____________

5 Add the following numbers to the place value chart.

a six hundred and fifty-eight

b forty-one thousand, one hundred and eighty

c fifteen thousand, seven hundred and nine

d nine thousand, one hundred and fifty-eight

	TTH	Th	H	T	O
a					
b					
c					
d					

Extension: On another sheet of paper, draw a number line and include each of the numbers from Questions 4 and 5.

Order of Operations

DATE:

1 Underline what needs to be worked out first. Then, complete the equations.

a $3 \times (3 + 3) =$

b $10 + 3 \times 8 =$

c $7 + 4 - 2 =$

d $20 - 4^2 =$

e $4 \times (3 + 8) =$

f $12 + 8 \div 4 =$

2 Place your own brackets in these equations to make them easier to solve. The first one has been done for you.

a $32 - 8 \times 3 + 1$

$= 32 - (8 \times 3) + 1$

$= 32 - 24 + 1$

$= 9$

b $14 \div 2 - 9 \div 3$

c $3^2 + 9 \times 5$

d $20 \div 4 + 36 \div 6$

e $8 \times 8 - 50 \div 5$

f $60 - 3 \times 8 \times 2$

3 Complete the following equations. Use brackets to help you work them out.

a $22 - 3 \times 3 + 4 \times 5$

b $9 \times 6 - 7^2 + 10 \div 2$

Extension: Create equations that include at least three signs for the following answers.

a 16

b 33

c 124

d 256

Emma's Farm

DATE:

Emma has taken over a farm just out of town. Even though it's not a huge farm, she still has a lot of work to do. Can you help her find out what she's got to work with?

1 There are 8 cows, 18 sheep and 10 chickens on Emma's farm. What is the total number of animal legs? ___________

2 One bag of feed can last 3 sheep a month. How many bags of feed will Emma need each year? ___________

3 Emma buys another 10 cows and 14 sheep. She splits all of her cows and sheep evenly into 2 paddocks. How many animals are in each paddock? ___________

4 Each of Emma's hens lays 8 eggs a month, except in winter, when they only lay 4 eggs a month. How many eggs will Emma get in her first year? ___________

5 There are 3 potato patches, 5 onion patches and 4 pumpkin patches. Each month, each potato patch grows 20 potatoes, each onion patch grows 25 onions and each pumpkin patch grows 12 pumpkins. How many vegetables will Emma have after one month? ___________

6 The vegetable patches are supposed to have 4 sprinklers each. Unfortunately, some sheep escaped from their paddock and damaged the pumpkin patch sprinklers. How many working sprinklers does Emma still have? ___________

Extension: Emma decides to sell 12 of her most troublesome sheep. When she bought the farm, they were worth $50 each.

a How much money will she make if she sells them now for $80 each? ___________

b With all the money from her sale, how many new (well-behaved) sheep could she buy at $50 each? ___________________________

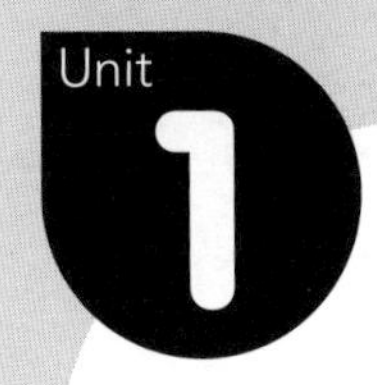

DATE:

STUDENT ASSESSMENT

1 State the value of the **5** in each number.

a 25 386 __________ **b** 91 053 __________ **c** 11 502 __________

2 Write the numbers in words.

a 2014 ______________________________

b 11 863 ______________________________

c 94 218 ______________________________

3 Answer the following.

a $2(3 \times 6) + (1 + 4) =$

b $8 \times 6 - (4 \times 3) =$

c $9 \div 3 + (8 \times 4) =$

d $(10 \times 10) \div (7 - 5) =$

e $42 - (3 \times 2) =$

f $(106 - 41) + (70 \div 7) =$

4 Nico has 20 boxes of apples. In each box there are 4 trays and each tray has 30 apples. How many apples does Nico have?

5 Jo collects trading cards. There are 4 subsets in each set. The total number of cards in each subset is 60.

a How many cards does Jo need to collect to have a complete set? __________

b If Jo has the following:

subset A: 51

subset B: 27

subset C: 49

subset D: 53

How many cards does Jo still need to collect? __________

All Four Operations

DATE:

1 Complete the addition equations.

a 38 + 16 =
b 142 + 26 =
c 268 + 243 =
d 1194 + 382 =
e 4238 + 3104 =
f 12572 + 4329 =

g 1068 + 735 = **h** 6143 + 1846 = **i** 42218 + 31405 + 1492 =

j At Friday night's football match, there were 78224 spectators. On Saturday, 86197 came to watch. 26216 watched Sunday's game. How many people went to the football on the weekend?

2 Complete the subtraction equations.

a 84 − 32 =
b 181 − 57 =
c 451 − 339 =
d 1240 − 618 =
e 6813 − 5158 =
f 14109 − 6163 =

g 3149 − 1428 = **h** 14283 − 10339 = **i** 63429 − 38753 − 10931 =

j A very lucky woman won $50000 in a raffle. With that, she bought a car for $24382, as well as a boat for $16788. How much does she have left?

3 Complete the multiplication tables.

a	3	5	9	12	15
× 6					

b	3	5	9	12	15
× 9					

4 Complete the multiplication equations.

a 18 × 6 = **b** 23 × 8 = **c** 38 × 9 = **d** 123 × 6 =

e Lachlan spends 9 hours a night sleeping. How much sleep does he get in 2 weeks?

5 Complete the division tables.

a	15	24	33	60	87
÷ 3					

b	24	40	64	96	128
÷ 8					

6 Complete the division equations.

a 48 ÷ 6 = **b** 99 ÷ 3 = **c** 184 ÷ 8 = **d** 231 ÷ 7 =

e Kaya shares 324 beads between herself and 8 friends. How many do they get each?

All Four Operations (TRB pp. 24–27)
Addition and subtraction MA3-5NA selects and applies appropriate strategies for addition and subtraction with counting numbers of any size
Multiplication and division MA3-6NA selects and applies appropriate strategies for multiplication and division, and applies the order of operations to calculations involving more than one operation

Missing Numbers

DATE:

1 Insert the missing number in each addition equation.

a $\begin{array}{r} 5\ 3 \\ +\square 5 \\ \hline 6\ 8 \end{array}$ **b** $\begin{array}{r} \square 9 \\ +\ 2\ 6 \\ \hline 1\ 1\ 5 \end{array}$ **c** $\begin{array}{r} 2\ 6\square \\ +\ 2\ 4\ 3 \\ \hline 5\ 0\ 8 \end{array}$ **d** $\begin{array}{r} 1\ 2\ 3\ 0 \\ +\ \square 8\ 4 \\ \hline 2\ 1\ 1\ 4 \end{array}$

e $384 + 1\square 5 = 529$ **f** $523\square + 3291 = 8528$ **g** $10102 + 122\square = 11331$

2 Insert the missing number in each subtraction equation.

a $\begin{array}{r} 8\ 7 \\ -\square 4 \\ \hline 5\ 3 \end{array}$ **b** $\begin{array}{r} 1\square 9 \\ -\ 2\ 6 \\ \hline 1\ 5\ 3 \end{array}$ **c** $\begin{array}{r} 3\ 1\square \\ -\ 1\ 8\ 7 \\ \hline 1\ 2\ 7 \end{array}$ **d** $\begin{array}{r} 1\ 0\ 7\ 3 \\ -\ 7\square 4 \\ \hline 3\ 1\ 9 \end{array}$

e $485 - 27\square = 213$ **f** $466\square - 1673 = 2990$ **g** $12683 - 9\square 85 = 3098$

3 Insert the missing number in each multiplication equation.

a $\begin{array}{r} 1\ 3 \\ \times\ \square \\ \hline 6\ 5 \end{array}$ **b** $\begin{array}{r} 1\square 3 \\ \times\ 6 \\ \hline 8\ 5\ 8 \end{array}$ **c** $\begin{array}{r} 3\ 4\square \\ \times\ 3 \\ \hline 1\ 0\ 2\ 6 \end{array}$ **d** $\begin{array}{r} 1530 \\ \times\ \square \\ \hline 12240 \end{array}$

e $38 \times \square = 266$ **f** $46\square \times 4 = 1876$ **g** $13\square 3 \times 5 = 6615$

4 Insert the missing number in each division equation.

a $\begin{array}{r} 1\ 7 \\ 3\overline{)\square 1} \end{array}$ **b** $\begin{array}{r} 2\ 2 \\ 6\overline{)13\square} \end{array}$ **c** $\begin{array}{r} 32 \\ 7\overline{)\square 24} \end{array}$ **d** $\begin{array}{r} 4\ 3 \\ 8\overline{)3\square 4} \end{array}$

e $1\square 4 \div 4 = 41$ **f** $31\square \div 7 = 45$ **g** $103\square \div 6 = 172$

5 Insert the missing operations and numbers, then complete the equations.

a $384 \ \square\ 98 = \square 86$ **b** $49\square\ \square\ 9 = 55$

c $1025\ \square\ 7 = \square 175$ **d** $5726\ \square\ 3\square 10 = 9536$

Extension:

a $12813\ \square\ 7392\ \square\ 2811 = 8232$ **b** $3100\ \square\ 6\square 0 = 5$

Fill the Gaps

DATE:

Work out the answer to each problem, then look for that number at the bottom of the page. Write the letter that goes with the problem above that number, to find the answer to the joke:
What do you call an alligator wearing a vest?

R	Renee gets \$8 pocket money each week. How many weeks until she can afford a rabbit (\$32) and a hutch for it to live in (\$120)? $8 \times$ ________ $= 32 + 120$
I	Ian has 6 green shirts, 11 white shirts and 9 blue shirts. If he has 30 shirts in total, how many shirts of other colours does he have?
T	Tamika has 14 strawberries, 19 blueberries and 16 mulberries. If she eats 7 berries each day, how many does she have left at the end of the week?
N	Natsuki has 9 friends. She wants to paint all of their fingernails, but not their thumbnails. If there is enough paint for 60 nails, how many nails will be left unpainted?
O	Omri has 15 oranges for his football team to eat at half time, which he has cut into quarters. How many quarters will he and 19 teammates each get?
S	Seth made 12 sandwiches. He cut half of them into halves, and the rest into quarters. How many sandwich slices does Seth have?
A	Akmal went on an adventure, where he found 6 rare coins worth \$3.50 each. When he went back, he found 2 more. How much could he sell all of them for?
E	Ellie has 28 earrings. She can wear 2 in each ear. If she wears different earrings each day, how many days will it be until she has worn them all?
I	Indy has 3 ice-cube trays with 14 blocks in each. She uses 3 in each glass of juice. If she drinks 11 glasses, how many ice blocks will be left?
G	Gavin picked 80 grapes from 4 different vines. He picked 32 grapes from the first vine, 18 from the second and 13 from the third. How many grapes did he pick from the last vine?
T	Tara's dog Blackie loves dog treats. There are 11 treats in each packet and Blackie eats 2 every day. How many packets does Tara need to buy in 30 days?
V	Vicky recorded a total of 3 minutes and 2 seconds of video. Each video clip was only 7 seconds long. How many separate video clips did she shoot?

An ____ ____ ____ ____ ____ ____ ____ ____ ____ ____ ____ ____

4 12 26 7 36 0 9 17 28 6 3 19

All Four Operations (TRB pp. 24–27)
Addition and subtraction MA3-5NA selects and applies appropriate strategies for addition and subtraction with counting numbers of any size
Multiplication and division MA3-6NA selects and applies appropriate strategies for multiplication and division, and applies the order of operations to calculations involving more than one operation

STUDENT ASSESSMENT

1 **a** $149 + 376$ **b** $456 - 287$ **c** 125×3 **d** $4\overline{)278}$

2 **a** $1069 + 2473 =$ **b** $8672 - 1146 =$

c $243 \times 7 =$ **d** $390 \div 6 =$

3 Find the missing numbers.

a $169 + 24\square = 416$ **b** $112\square - 347 = 781$ **c** $14 \times \square = 84$ **d** $8\overline{)15\square}$ = 19

4 Find the missing numbers.

a $4287 + 11\square8 = 5455$ **b** $7932 - 4\square63 = 3369$

c $1\square6 \times 9 = 1314$ **d** $1984 \div \square = 248$

5 Write a number sentence for each problem, and solve them.

a Divide a number by 12, then add 9 to get 16.

b Increase a number by 78, then multiply by 2 to get 166.

c Double a number, then multiply by 5 to get 720.

Extension:

a $68 \div 3 =$ **b** $92 \div 8 =$ **c** $156 \div 5 =$ **d** $350 \div 9 =$

e Jun gave an equal number of his 155 pencils to his 6 friends. He kept the pencils that were left over. How many pencils did each of his friends get? How many were left for Jun?

Knights in Maths

DATE:

Why did the knight run about shouting for a tin opener? Match the numbers and words, and then write the letters below to find the answer. The first one has been done for you.

1	negative eight	**H**	4	four	**I**
-3	two	**E**	3	negative nine	**F**
5	ten	**B**	7	negative six	**T**
-8	negative three	**N**	-6	seven	**S**
-10	zero	**U**	-9	three	**R**
2	negative ten	**D**	-1	eight	**Y**
0	five	**A**	8	negative one	**M**
10	one	**O**	-2	negative two	**W**

__ __   __ __ __   __   __ __ __
-8 2   -8 5 -10   5   10 2 2

__ __   __ __ __   __ __ __ __
4 -3   -8 4 7   7 0 4 -6

O __   __ __ __   **O** __ __
1 -9   5 3 -1   1 0 3

Extension: On another sheet of paper, create your own puzzle.

Integers (TRB pp. 28–31)
Whole numbers MA3-4NA orders, reads and represents integers of any size and describes properties of whole numbers

Ordering Integers

DATE:

1 Order each set of numbers from **smallest** to **largest**.

a 2, 3, 5, 0, –1, –2 ______

b 2, 4, 0, –2, 6, –4 ______

c –1, –3, 5, 0, 3, –5, 1 ______

d 19, 14, 0, –10, 15, –9, –13 ______

2 Show the following numbers on a number line.

a 2, –3, 0, 5, –4, 3

b –8, 7, –6, 5, 0, 1, 4

c 3, –10, 5, –8, –6, 0, 2

d –9, 5, 7, 8, –6, –4, –2

3 Circle the **largest** number in each pair.

a –2 6

b 5 2

c –3 –1

d –7 –8

e 2 –5

f 2 –5

g –2 3

h 0 –4

4 Describe how you worked out which number was largest in Question 3e.

______

Extension: Using all of the digits 5, 6, 2, 1:

a Write the **smallest** number. ______

b Write the **largest** number. ______

c What might the **smallest** negative number be? ______

Addition with Negative Numbers

DATE:

1 During winter, the temperature was 5° C. What would the temperature be if it was:

a 4 degrees warmer? ___________ **b** 5 degrees cooler? ___________

c 7 degrees warmer? ___________ **d** 10 degrees cooler? ___________

50
40
30
20
10
- 0
-10
-20
-30
-40
°C

2 Use the number line to solve each equation.

a 10 + -6 =

b 2 + -5 =

c 3 + -8 =

d -2 + -3 =

3 Complete each equation.

a 6 + -4 = **b** -7 + -8 = **c** -2 + 3 =

d 2 + -5 = **e** 0 + -4 = **f** 0 + -4 =

g 5 + 2 = **h** 2 + -5 =

4 Describe any patterns you notice when adding negative numbers.

____

Extension: Use the graph to describe what is happening to the temperature during the day.

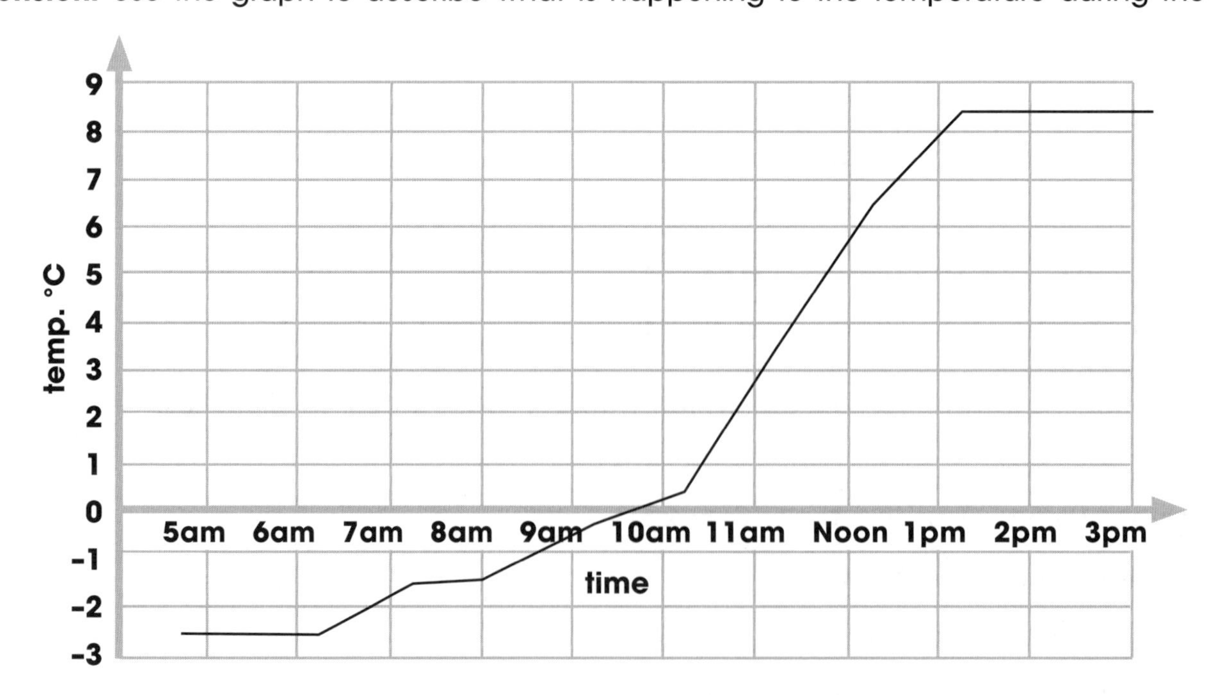

____

DATE:

STUDENT ASSESSMENT

1 Draw lines to match the numbers and words.

-3	negative six
15	negative eighteen
0	twenty-one
-18	negative three
21	zero
-6	fifteen

2 Draw each set of numbers on a number line.

a -5, 2, 0, 4, 8, -3

b -7, -1, 0, 5, -2, 8, 4

3 Circle the **largest** number in each pair.

a 3 -5 **b** 1 0 **c** -7 -5 **d** 9 11

4 Use a number line to solve each equation.

a -7 + -3 =

b 4 + -5 =

c 0 - 4 =

d -1 - 3 =

5 Complete the equations.

a 3 + -4 = **b** -5 + 3 = **c** -1 + -3 = **d** 6 + -7 =

6 Circle the **largest** number: 42 631 42 813

Describe how you worked this out. ____________________

Perimeter

DATE:

Work out the perimeter of each shape.

a

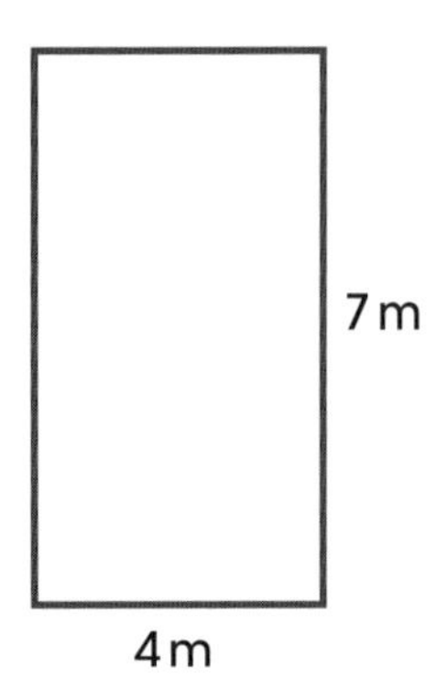

P = ________

b

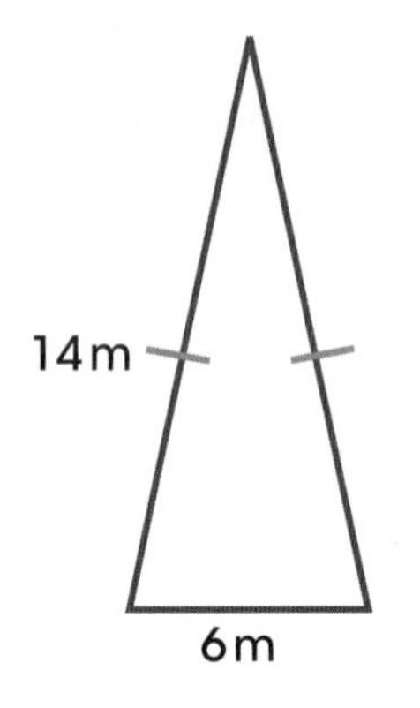

P = ________

c

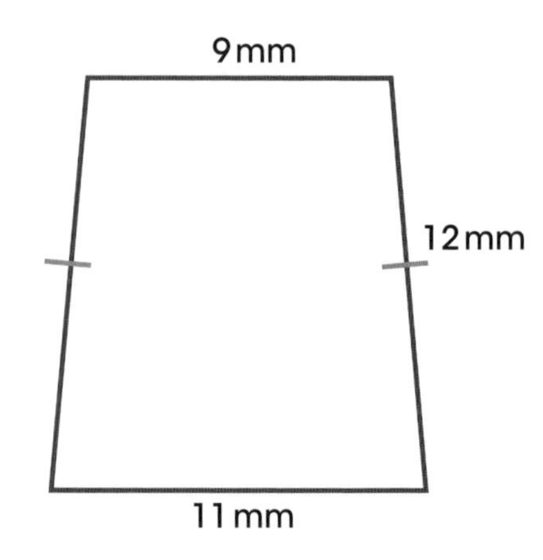

P = ________

d

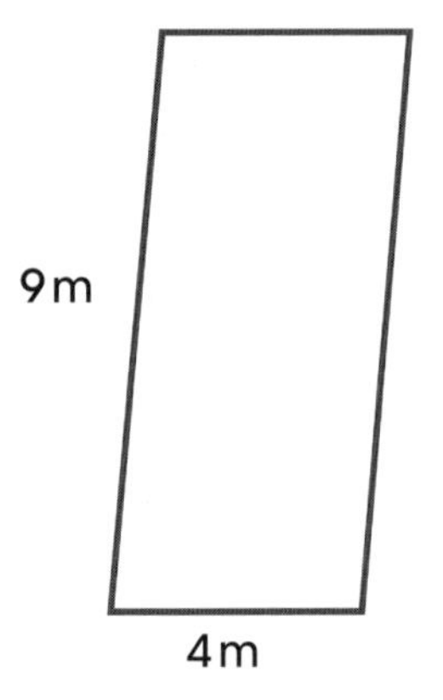

P = ________

e

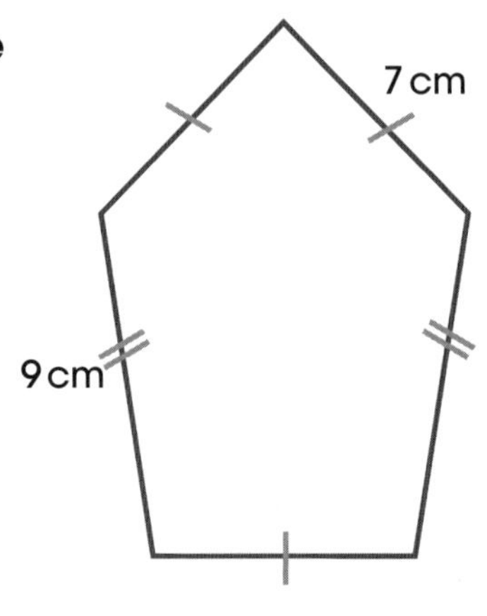

P = ________

f

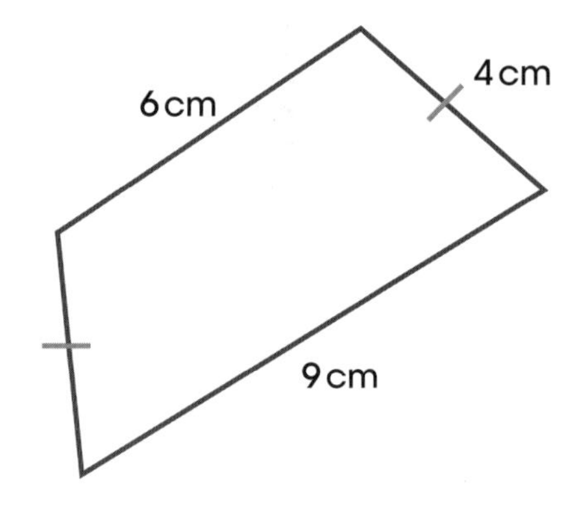

P = ________

g

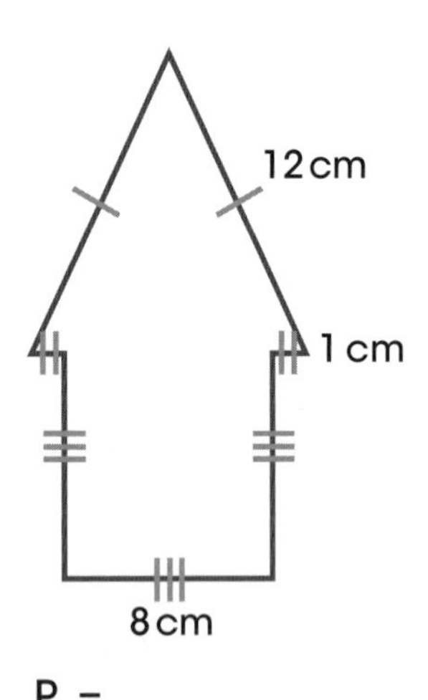

P = ________

h

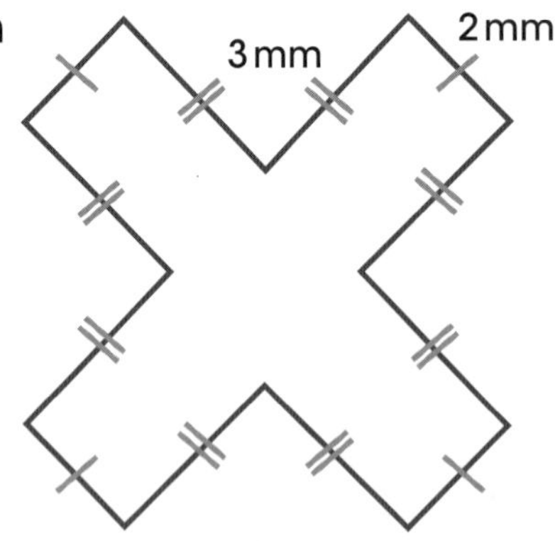

P = ________

i

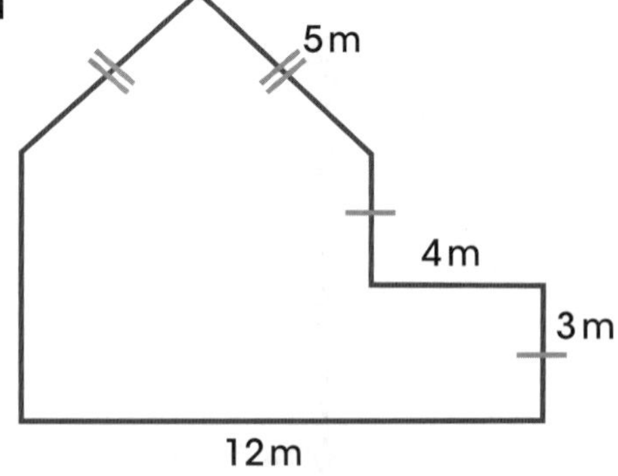

P = ________

Extension: On another sheet of paper, draw your own shape.
It must have at least 6 sides, and the perimeter must add up to 30 cm.

Area

DATE:

1 Find the area of each square.

a

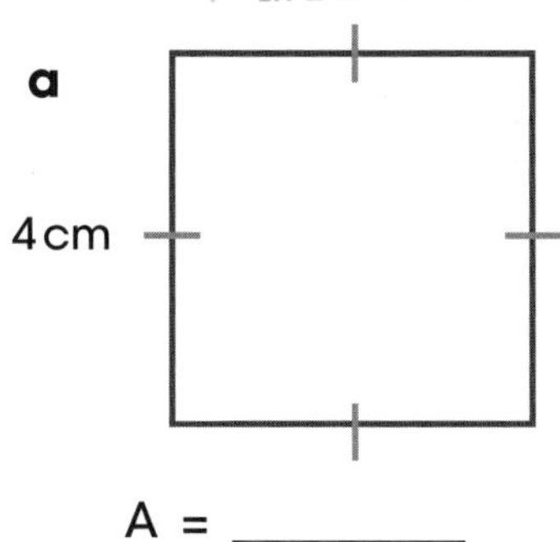

A = ________

b

2 mm

A = ________

c

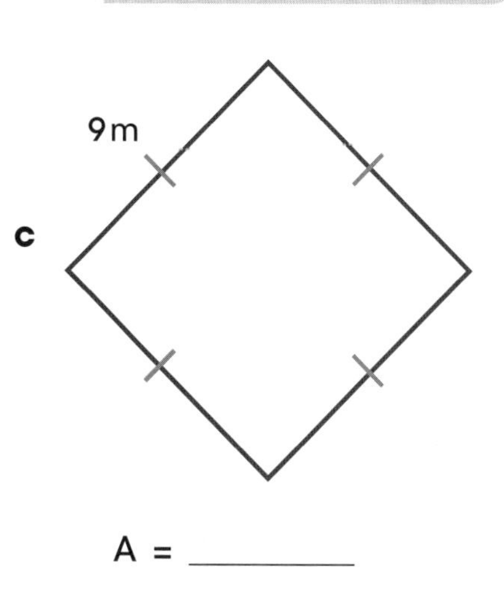

A = ________

d What do you notice about the area of squares?

____

2 Find the area of each rectangle.

a

5 mm

11 mm

A = ________

b

9 cm

2 cm

A = ________

c

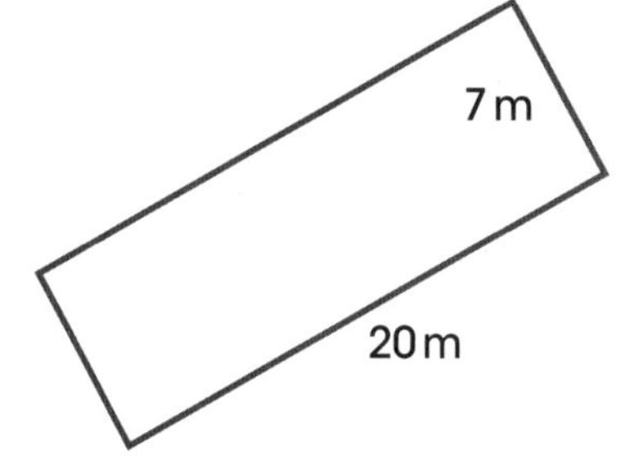

A = ________

3 Find the area of each triangle.

a

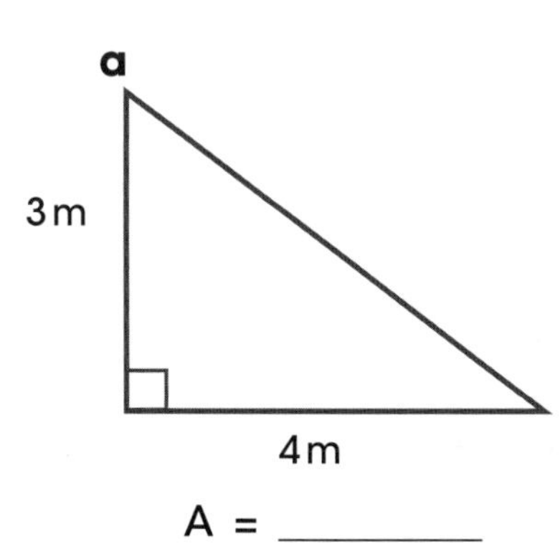

A = ________

b

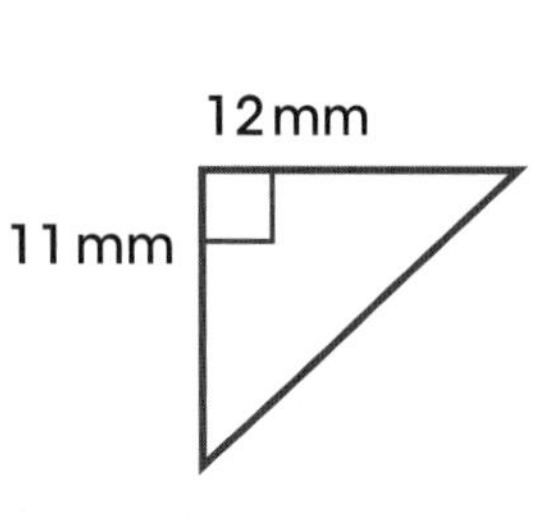

A = ________

c

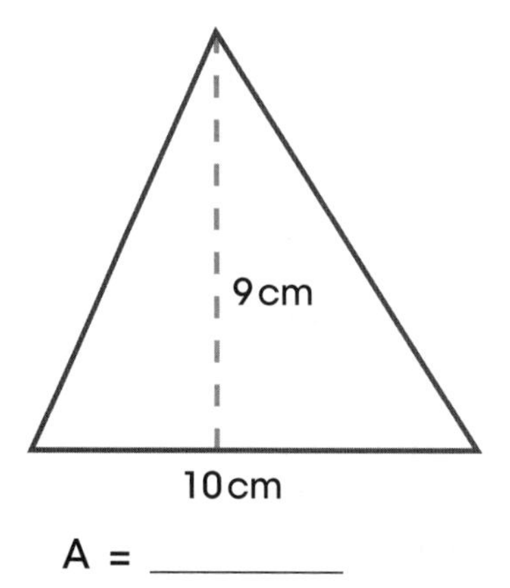

A = ________

d How does the area of a triangle relate to the area of a square/rectangle? Explain using an example.

____

____

Extension: On another sheet of paper, draw your own shape.
It must have at least 3 sides, and the area must equal 30 cm^2.

Area and Perimeter

DATE:

Find the perimeter and area of each shape.

a

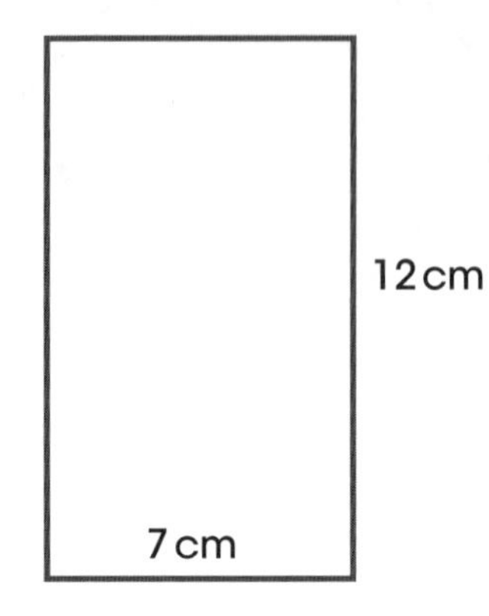

P = ________

A = ________

b

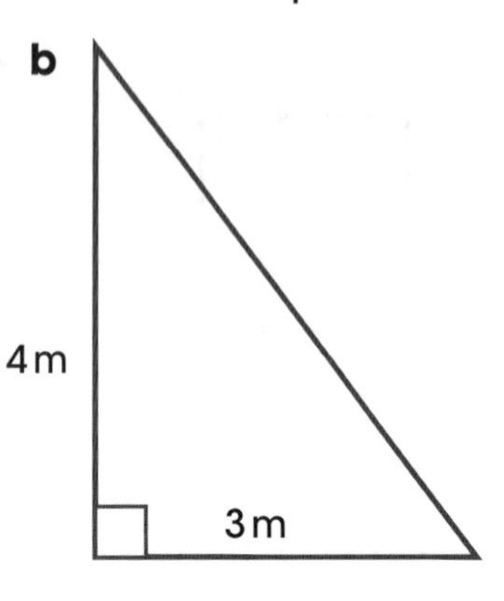

P = ________

A = ________

c

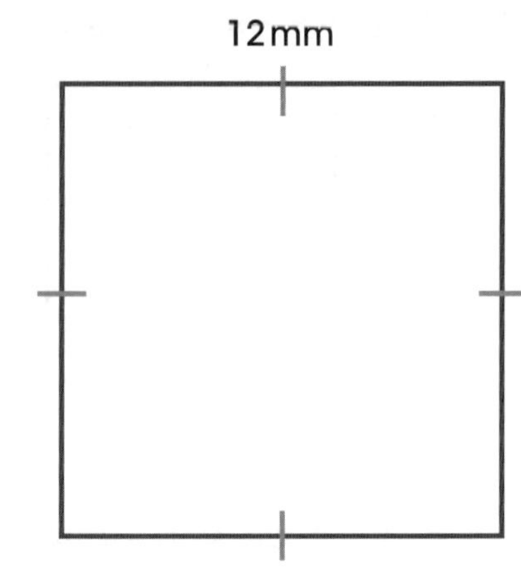

P = ________

A = ________

d

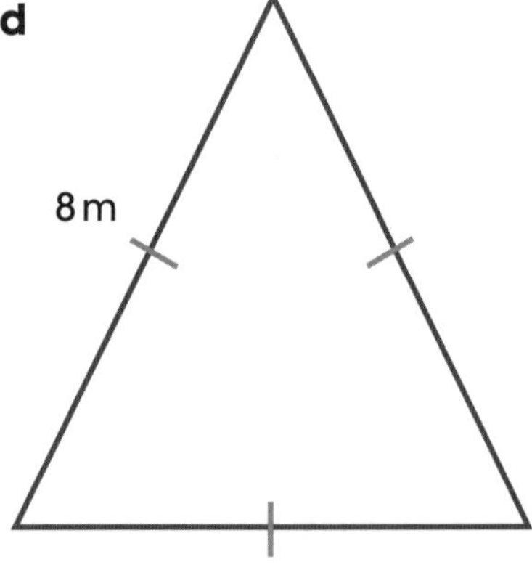

P = ________

A = ________

e

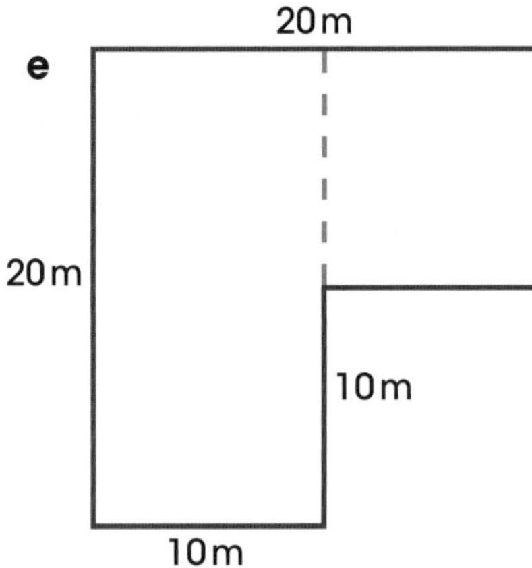

P = ________

A = ________

f

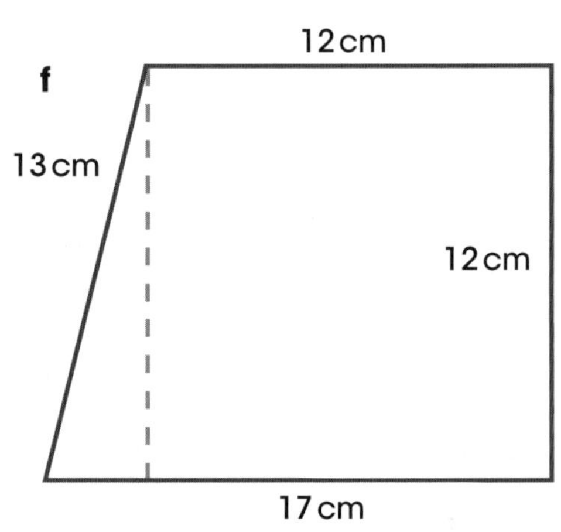

P = ________

A = ________

g

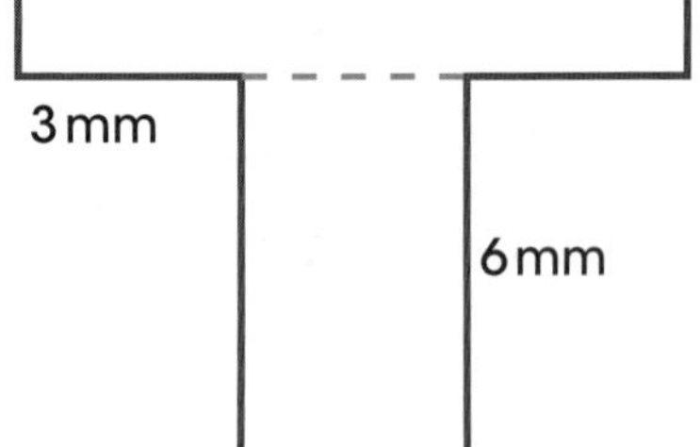

P = ________

A = ________

h

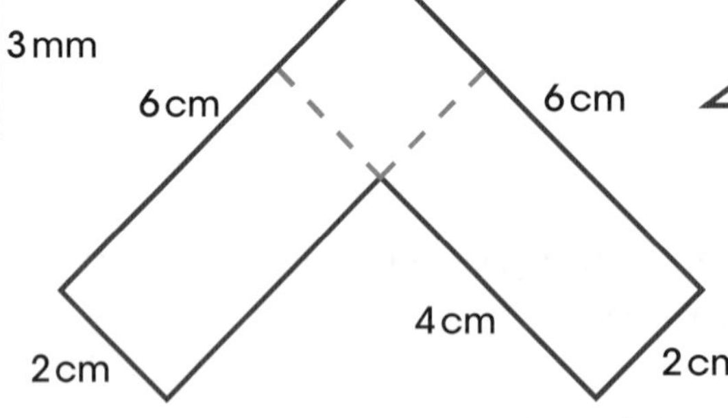

P = ________

A = ________

i

P = ________

A = ________

Area and Perimeter (TRB pp. 32–35)
Length MA3-9MG selects and uses the appropriate unit and device to measure lengths and distances, calculates perimeters, and converts between units of length
Area MA3-10MG selects and uses the appropriate unit to calculate areas, including areas of squares, rectangles and triangles

DATE:

STUDENT ASSESSMENT

1 Define each term.

a length: ___

b perimeter: ___

c area: ___

2 List two units each that are used to measure:

a perimeter: ___ **b** area: ___

3 Find the perimeter of each shape.

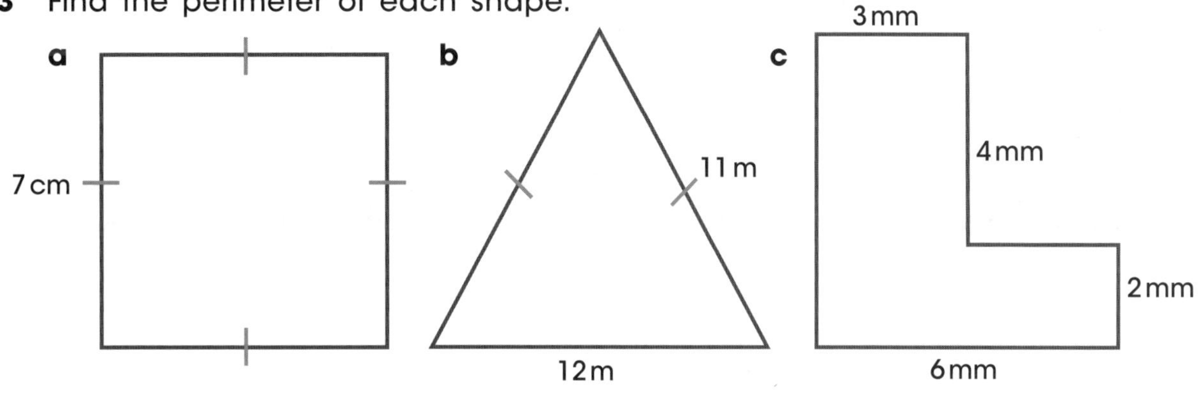

P = ___ P = ___ P = ___

4 Find the area of each shape.

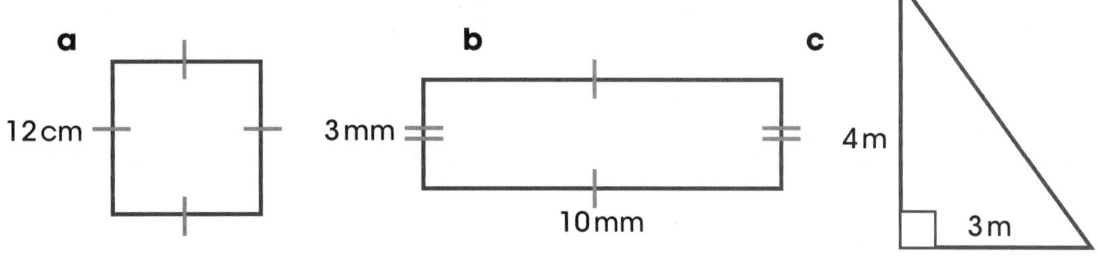

A = ___ A = ___ A = ___

5 Find the area of each composite shape.

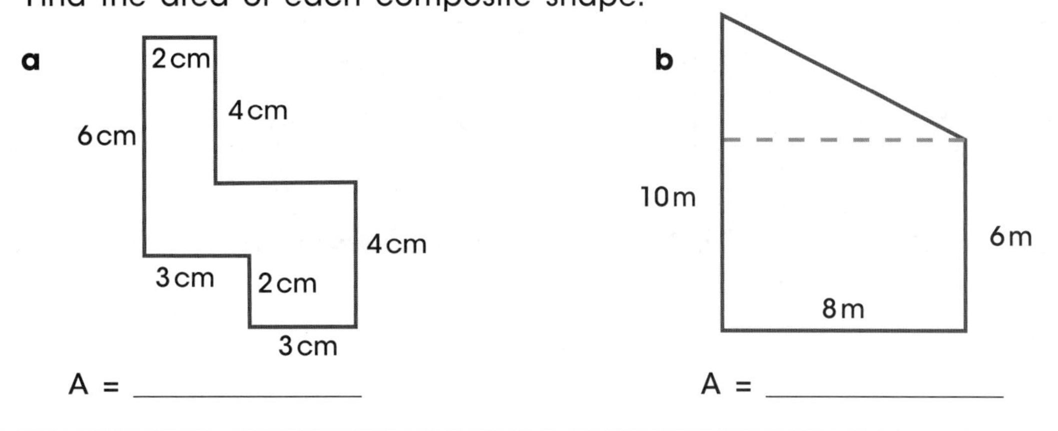

A = ___ A = ___

Reading Scales

DATE:

1 Read the scales and write the answers.

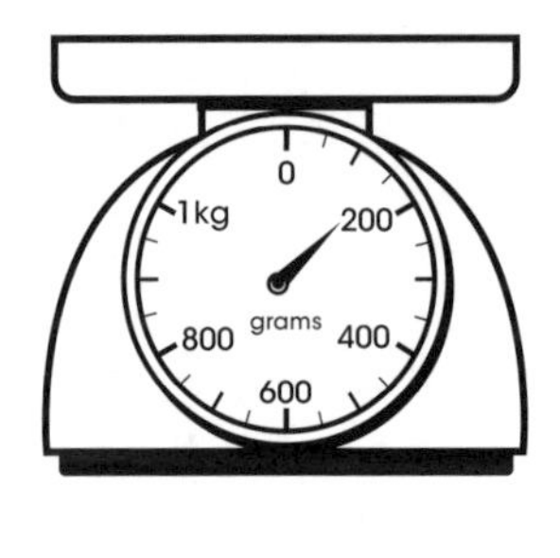

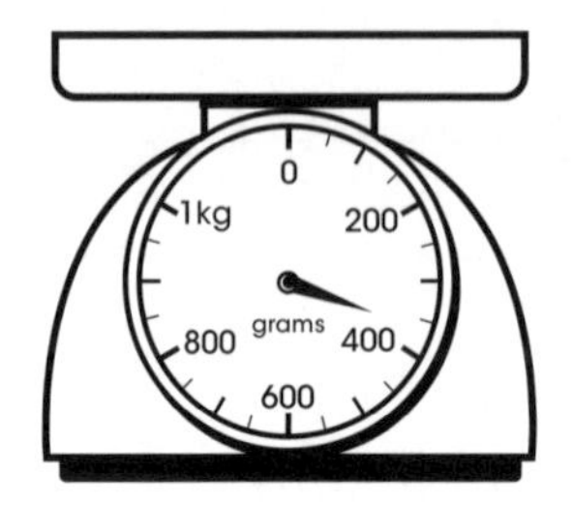

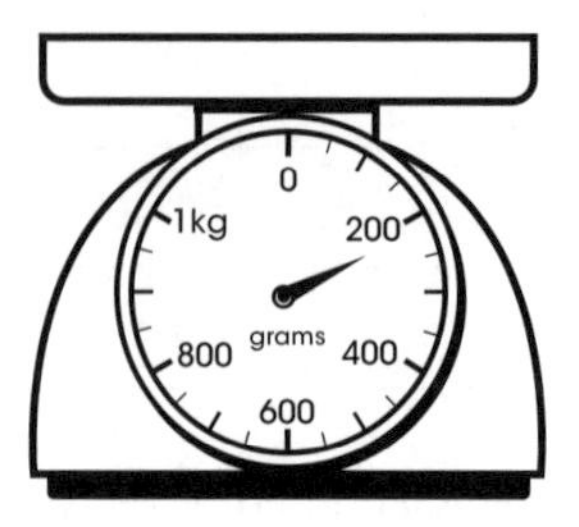

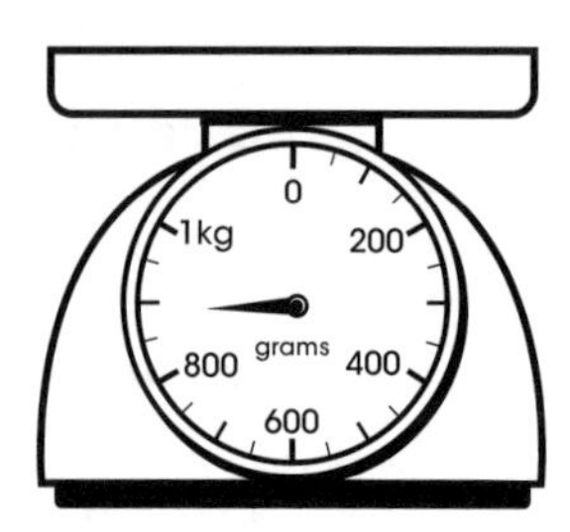

a ________ g **b** ________ g **c** ________ g **d** ________ g

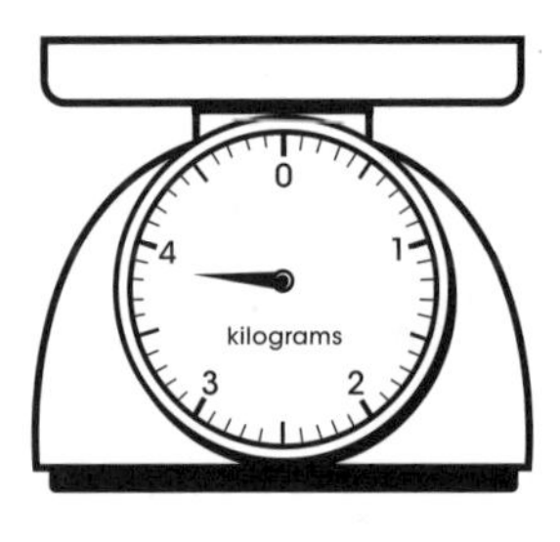

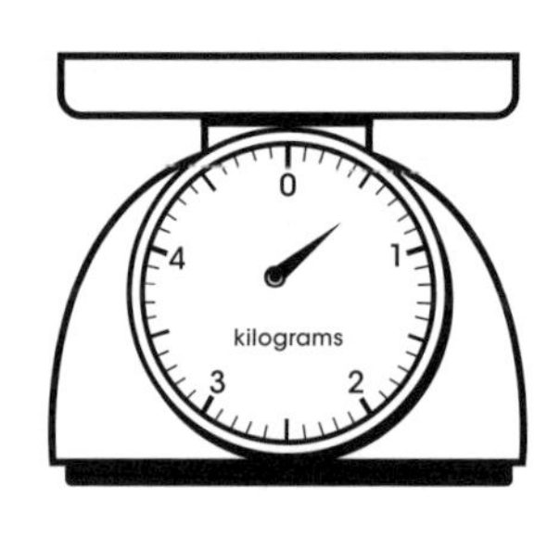

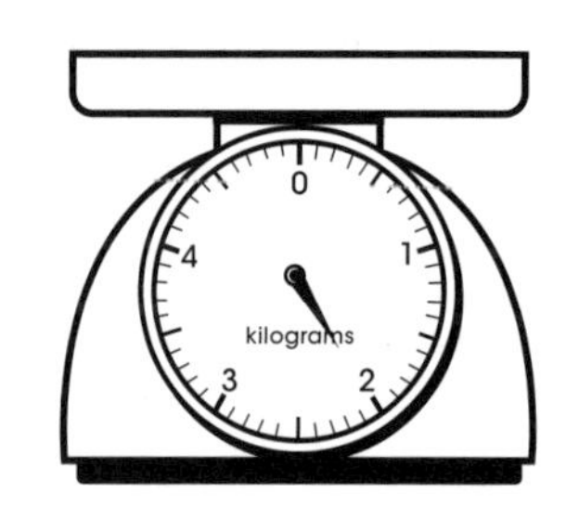

e ________ kg **f** ________ kg **g** ________ kg **h** ________ kg

2 Read the scales and write the answers in both grams and kilograms.

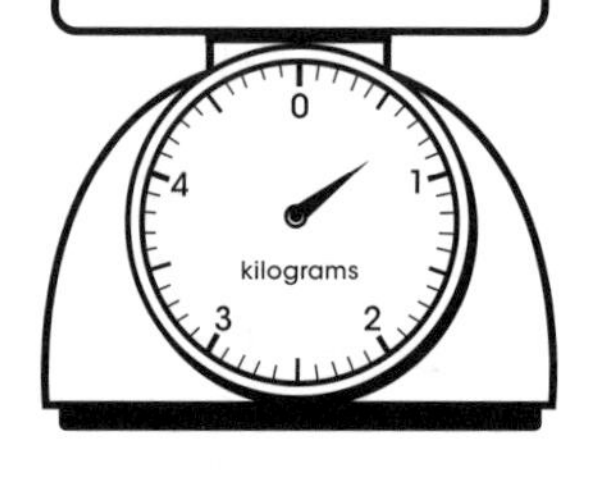

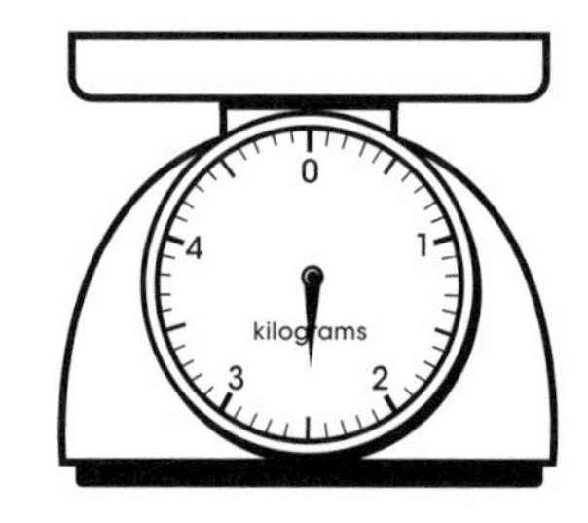

a ________ g **b** ________ g **c** ________ g **d** ________ g

________ kg ________ kg ________ kg ________ kg

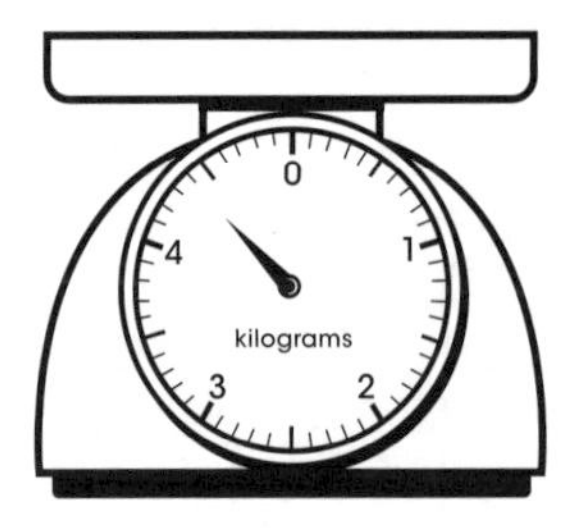

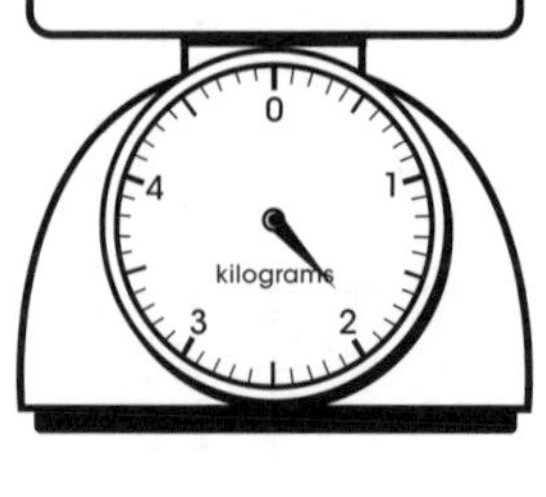

e ________ g **f** ________ g **g** ________ g **h** ________ g

________ kg ________ kg ________ kg ________ kg

Capacity and Volume

DATE:

You will need: coloured pencils

1 Tick the most appropriate unit (mL or L) for measuring the capacity of the following.

	Object	mL	L
a	Coffee cup		
b	Bucket		
c	Medicine cup		
d	Swimming pool		
e	Fish tank		
f	Drinking glass		

2 Shade the jugs to show the given capacity.

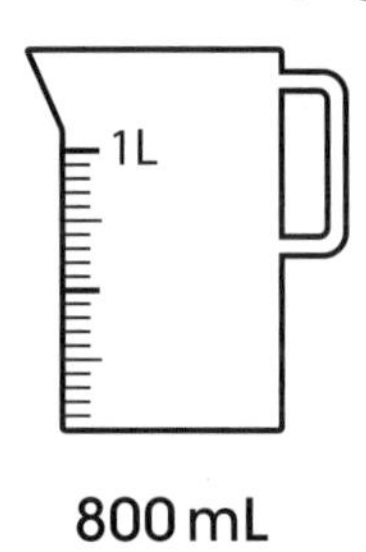

800 mL

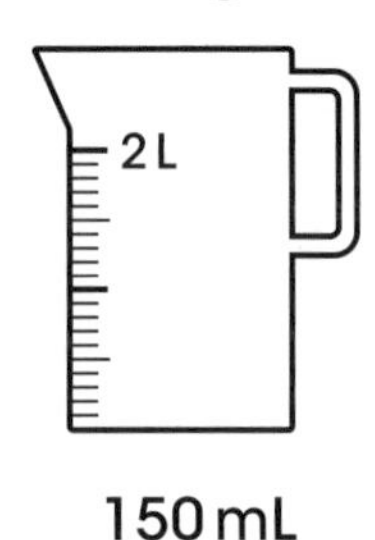

150 mL

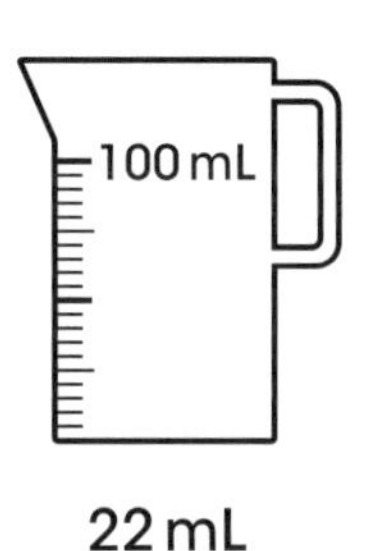

22 mL

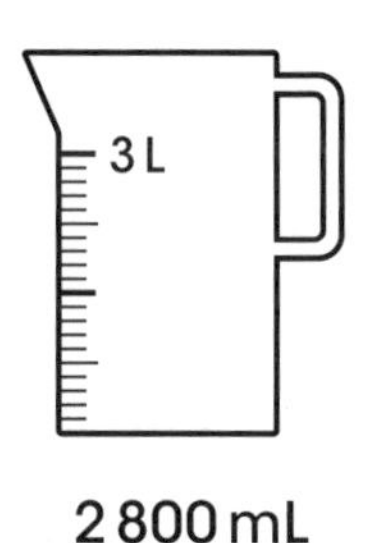

2800 mL

3 State the volume of each shape.

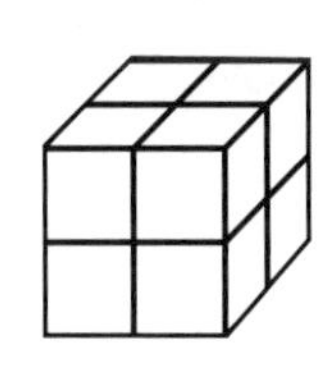

V = ________

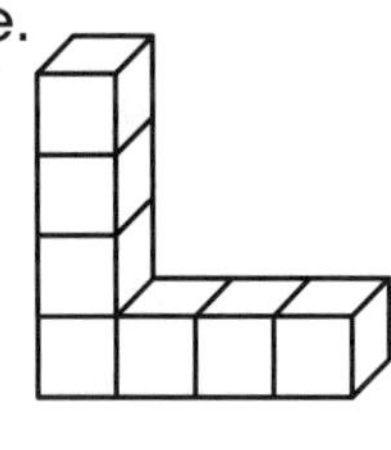

V = ________

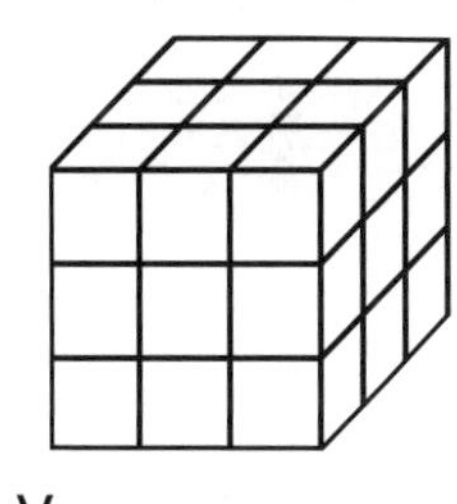

V = ________

Extension:

a How might you separate 2 L of water into six different sized containers? Draw your answer in one of the boxes below.

b Draw a shape like the shapes in Question 3 with a volume of 29 cubic units.

Units of Mass, Capacity and Volume

DATE:

1 Express each of the following in litres.

a 4250 mL = ________ **b** 15000 mL = ________

c 800 mL = ________ **d** 70 mL = ________

2 Express the following in litres, using decimal notation.

a 6 L 400 mL = ________ **b** 25 L 250 mL = ________

c 31 258 mL = ________ **d** 48 456 mL = ________

3 Express the following in kilograms, using decimal notation if necessary.

a 6500 g = ________ **b** 15289 g = ________

c 58 360 g = ________ **d** 5 kg 900 g = ________

e 10 kg 200 g = ________ **f** 15 kg 15 g = ________

4 Complete the table for measurements of water.

	Capacity	Volume	Mass
a	10 mL		
b		400 cm^3	
c			900 g
d			5 kg
e		850 cm^3	
f	1.7 L		

Extension: 1 cm^3 of water = 1 mL = 1 g.

How is this different for a substance such as sand? Investigate.

Mass and Capacity (TRB pp. 36–39)
Volume and capacity MA3-11MG selects and uses the appropriate unit to estimate, measure and calculate volumes and capacities, and converts between units of capacity
Mass MA3-12MG selects and uses the appropriate unit and device to measure the masses of objects, and converts between units of mass

DATE:

STUDENT ASSESSMENT

1 Complete the table.

	Definition	Units
Mass		
Capacity		
Volume		

2 Express each mass in kilograms.

a 200 g ____________ **b** 1 600 g ____________ **c** 2 500 g ____________

3 State the volume of each object.

a

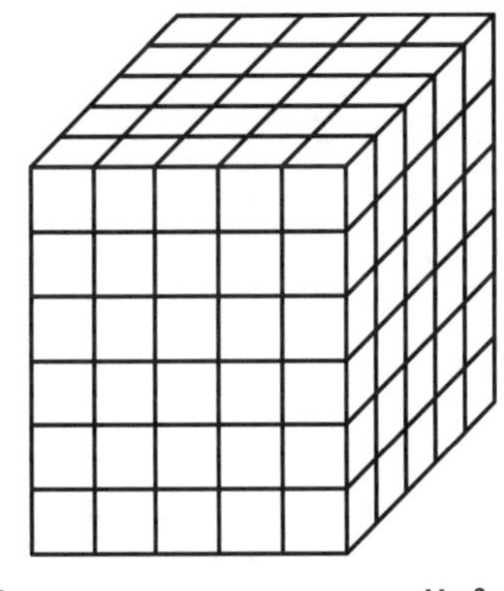

V = ____________ units³

b

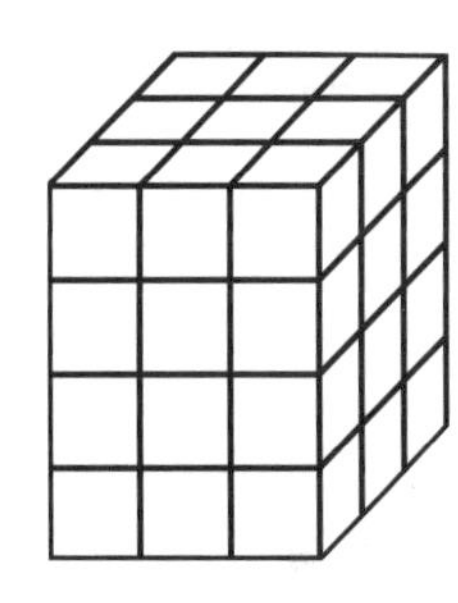

V = ____________ units³

4 State the amount of water in each container.

a

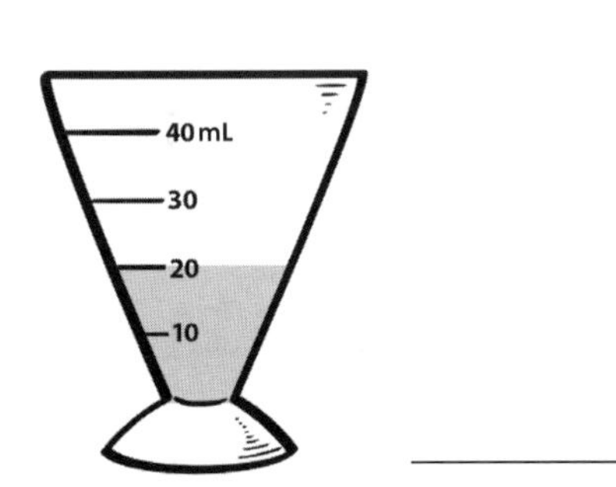

____

b

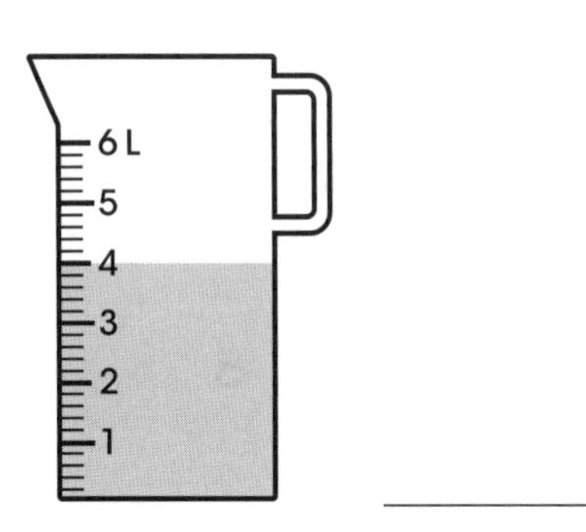

____

5 Complete the table for measurements of water.

500 mL	L
1 500 cm³	mL
2 500 g	kg

Find the Odd Numbers

DATE:

You will need: coloured pencils

Work out the answer to each of the 30 questions. Then, find and colour the answer in the number search below. When you've finished, you will see the answer to the question: **Are all prime numbers odd?**

15 + 8 =	49 ÷ 7 =	3 × 7 =	11 × 5 =	66 ÷ 2 =
13 × 3 =	$(9^2) - 10 =$	66 + 25 =	66 ÷ 22 =	77 − 18 =
9 × 7 =	54 ÷ 54 =	9 × 3 =	$7^2 =$	33 − 16 =
$3^2 =$	17 × 3 =	33 × 3 =	7 × 5 =	55 ÷ 11 =
$9^2 =$	33 + 56 =	45 ÷ 3 =	29 × 3 =	44 + 23 =
19 × 3 =	15 + 16 =	77 ÷ 7 =	125 ÷ 5 =	$(6^2) + 1 =$

31 45 69 3 95 73 71 51 11 75 53 13 59

63 23 85 35 99 29 47 9 97 87

39

89 69 27 79 81 91 65 95 77 15 73

55

49

57 19 5 37 41 13 61 21 75 13

25 43 93 17 83 33 7 67 85 1

Extension: Find a pattern of numbers in the number search that looks like the first letter of your name, and colour those numbers. Use a different colour.
On another sheet of paper, create your own questions for the numbers you've coloured.

Prime Numbers

DATE:

You will need: coloured pencils

1 Colour in all the prime numbers on the chart.

1	2	3	4	5	6	7	8	9	10
11	12	13	14	15	16	17	18	19	20
21	22	23	24	25	26	27	28	29	30
31	32	33	34	35	36	37	38	39	40
41	42	43	44	45	46	47	48	49	50
51	52	53	54	55	56	57	58	59	60
61	62	63	64	65	66	67	68	69	70
71	72	73	74	75	76	77	78	79	80
81	82	83	84	85	86	87	88	89	90
91	92	93	94	95	96	97	98	99	100

2 How did you know 73 was a prime number?

____

3 Can you see a pattern in the chart? If so, describe it.

____

4 Is 2 a prime number? Explain your answer.

____

Extension: Are the following numbers prime? Explain your answers.

101 **143** **149** **333**

____

Prime Number Investigation

DATE:

1 Write down the current year: ___________

Is it a prime number? Explain how you know.

____

____

2 Not including this year, what is the next year that will be a prime number? ___________

3 Can you think of a good way to work out if a number is prime? This can include using calculators or other technology.

____

____

4 Are the following years prime? Circle the correct response.

Your birth year: ___________	Yes / No
The year you started school: ___________	Yes / No
The year Australia became an independent nation: 1901	Yes / No
The year the Victorian Gold Rush began: 1851	Yes / No
The year the First Fleet set sail from England: 1787	Yes / No
The year you will start high school.	Yes / No
The year one of your parents or your carer was born.	Yes / No

Extension: Challenge your friends by giving them four-digit numbers. Ask them to work out if the number is prime or not.

Record the numbers you gave your friends, and whether they were prime.

a ____

b ____

c ____

d ____

STUDENT ASSESSMENT

1 Circle the odd numbers.

20 11 101 596 2087

2 Circle the even numbers.

56 70 403 1076 3178

3 What is a prime number? ______________________________

____

4 Are all prime numbers odd? Explain. ______________________________

____

5 Circle the prime numbers.

1 2 4 7 11 19 21 30 45 61

6 List the factors of each number.

a 16 ______________________________

b 23 ______________________________

c 47 ______________________________

d 52 ______________________________

e 101 ______________________________

7 Which numbers from Question 6 are prime numbers? ______________________________

____

8 Describe how to find factors for a number. Use the number 20 as an example.

____

____

____

Composite Numbers

DATE:

1 Cross out the prime numbers.

2 5 7 9 11 17 20 19 23 27 30

2 Create factor trees for each of the composite numbers above.

3 Draw factor trees for each number.

28 

36 

40

48

4 Fill in the gaps in the factor trees.

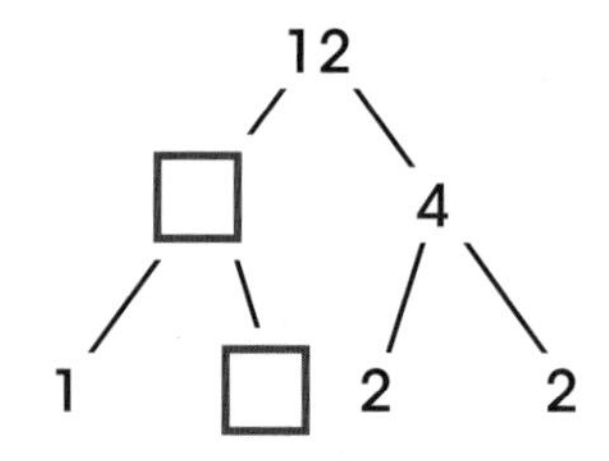

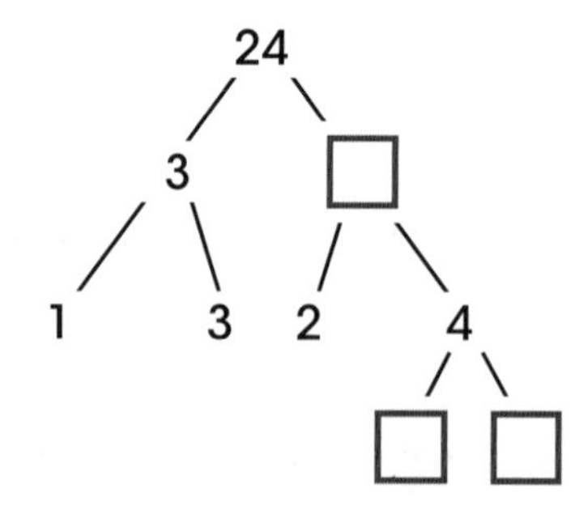

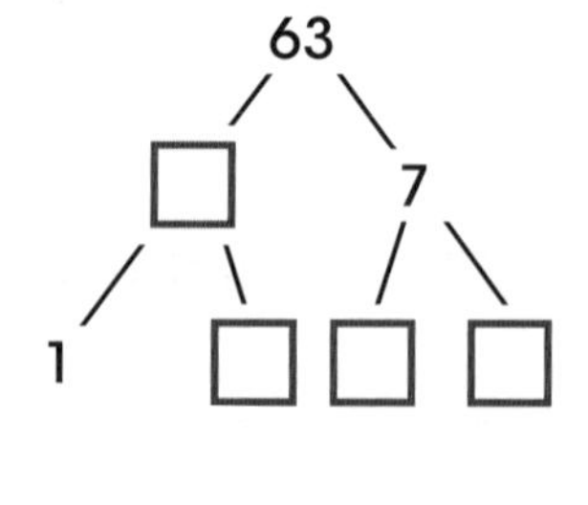

Extension: On another sheet of paper, draw factor trees for each of these numbers.

96 144 160 216 250

Using the Rules of Divisibility

DATE:

1 Explore each of the rules of divisibility with the examples below. Circle the numbers that are divisible by the number in the first column.

Divisible by	Rule	Numbers			
2	Last digit is divisible by 2.	459	720	456	889
3	The sum of the digits is divisible by 3.	156	7 861	1 079	4 567
4	The last 2 digits make a number that is divisible by 4.	179	1 058	4 462	33 980
6	The number is divisible by 2 and 3.	48	196	3 284	10 980
8	The last 3 digits make a number that is divisible by 8.	3 616	1 709	44 826	11 284
9	The sum of the digits is divisible by 9.	1 169	1 872	48 935	110 872

2 Use the rules of divisibility to find the prime factors of each number.

a 89 ____________________

b 167 ____________________

c 872 ____________________

d 1 042 ____________________

e 3 846 ____________________

Square and Triangular Numbers

DATE:

1 Circle the square numbers. Underneath each square number, write its square root. One has been done.

16	25	33	36	42	49	56	64
4	______	______	______	______	______	______	______

2 Using the triangle as a guide, list the first ten triangular numbers.

1

3

6

______

______

______

______

______

______

______

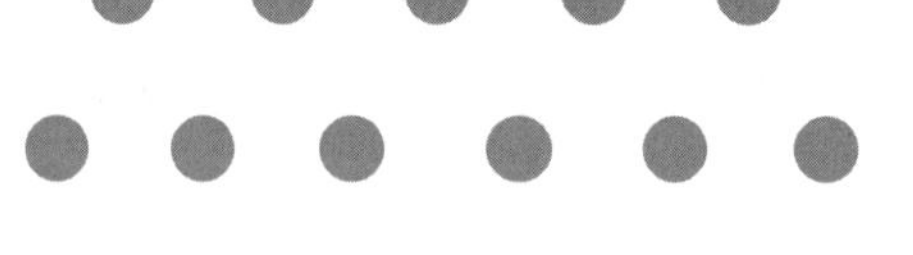

3 Does this triangle still show triangular numbers?

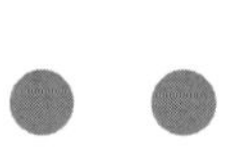

____

Extension: On another sheet of paper, make your own number pattern based on a shape. For example, diamond numbers or hexagonal numbers.

DATE:

STUDENT ASSESSMENT

1 Draw factor trees for each number.

60 

42 

2 List all of the factors for each number.

a 36 ______________________

b 48 ______________________

c 102 ______________________

d 156 ______________________

3 What is a composite number?

____

Give two examples: **a** ____________ **b** ____________

4 Give the prime factors of your composite numbers from Question 3.

a ______________________

b ______________________

5 Is 100 a square number? Explain. ______________________

____

6 Is 100 a triangular number? Explain. ______________________

____

Polygons

DATE:

1 Name each polygon.

a 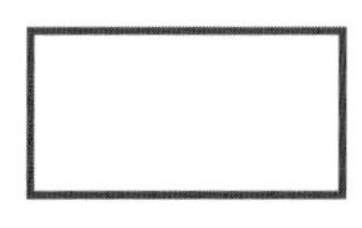

b 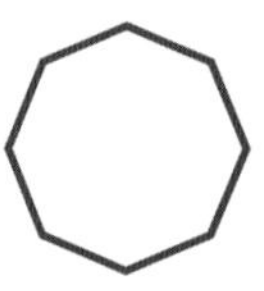

c 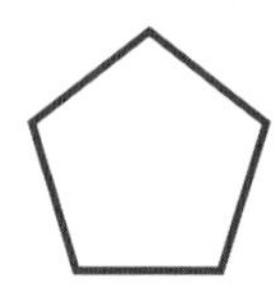

d 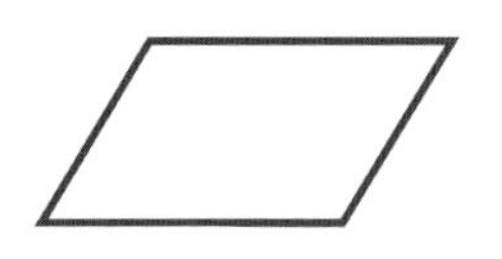

e 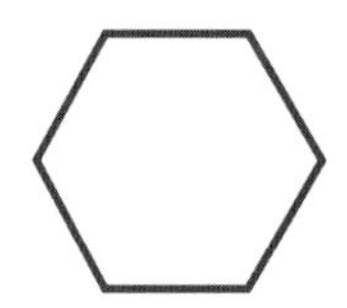

f 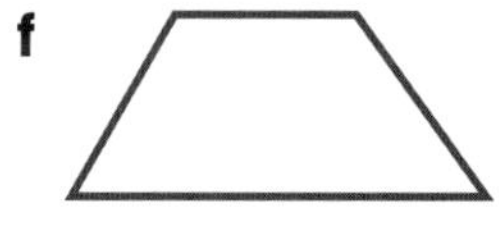

2 Which of the polygons in Question 1 have the same number of diagonals? Shade one of the circles below.

○ **a, b** and **c** ○ **a, d** and **e** ○ **a, d** and **f** ○ **a, c** and **f**

3 If you had to count the number of diagonals on each shape in Question 1, which shape would be the most difficult? Why? (Try drawing them.)

4 There is a way to find the number of diagonals in a polygon without drawing and counting them. To do this, you need to know the formula: $n \times (n - 3) \div 2$. The letter n stands for the number of sides in the polygon. You also need to know the order of operations, sometimes called BODMAS (see Unit 1, p.5). Here is how it works for a square.

Step 1: Replace n with the number of sides. $4 \times (4 - 3) \div 2$

Step 2: Work out the brackets first. $4 \times 1 \div 2$

Step 3: Work out the rest of the number sentence. $4 \div 2 = 2$

The formula works. The number of diagonals in a square is 2. 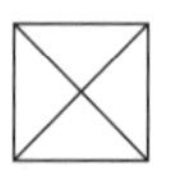

On a separate sheet of paper, test the formula for other polygons, starting with a pentagon.

2D Shapes (TRB pp. 48–51)
Two-dimensional space MA3-15MG manipulates, classifies and draws two-dimensional shapes, including equilateral, isosceles and scalene triangles, and describes their properties

Note: This unit focuses on two-dimensional shapes, including those that make up the faces of three-dimensional objects, before students move on to the specific study of prisms and pyramids in Unit 9.

Circles

DATE:

You will need: a drawing compass, A4 paper

1 Use the words below to label the circles.

diameter circumference semi-circle sector radius quadrant

The arrow is pointing to the ____________	The arrow is pointing to the ____________	The arrow is pointing to the ____________
The shaded part is a ____________	The shaded part is a ____________	The shaded part is a ____________

2 **a** Put a dot in the centre of a sheet of A4 paper.

b Set your compass at 8 cm and draw a circle with the dot as the centre.

c What is the diameter of the circle? ____________

d Use the same centre point to draw three more circles with the following diameters:

14 cm 12 cm 10 cm

3 Use your compass to create some circle patterns on a sheet of A4 paper.

2D Shapes (TRB pp. 48–51)
Two-dimensional space MA3-15MG manipulates, classifies and draws two-dimensional shapes, including equilateral, isosceles and scalene triangles, and describes their properties

Note: This unit focuses on two-dimensional shapes, including those that make up the faces of three-dimensional objects, before students move on to the specific study of prisms and pyramids in Unit 9.

Nets

DATE:

1 Name the solid that matches each net.

a

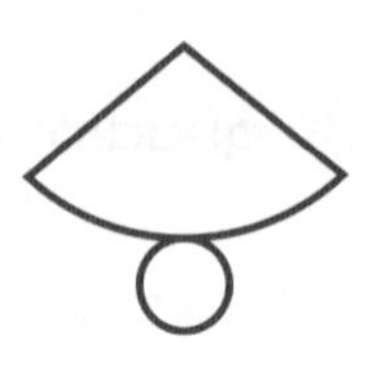

b

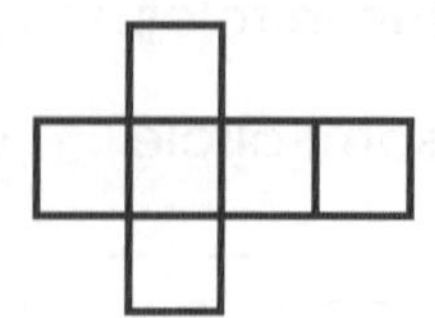

c

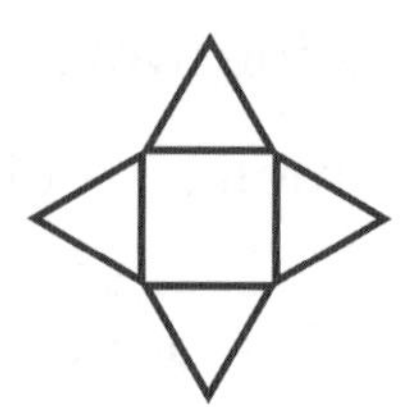

d

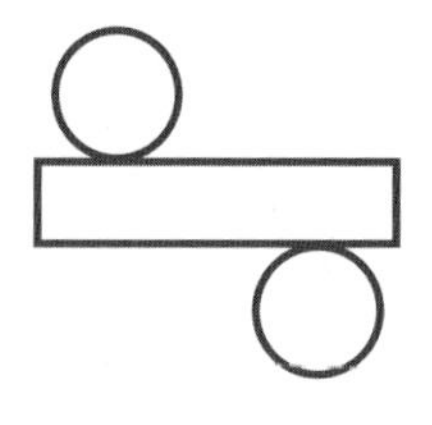

e

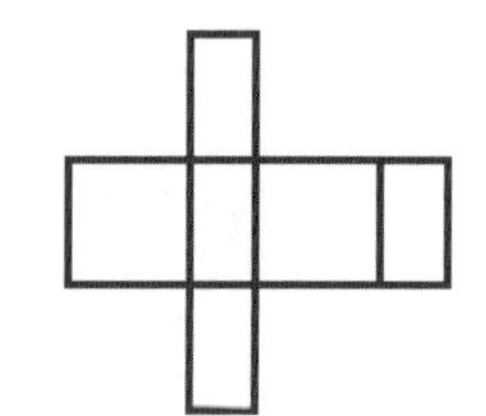

f

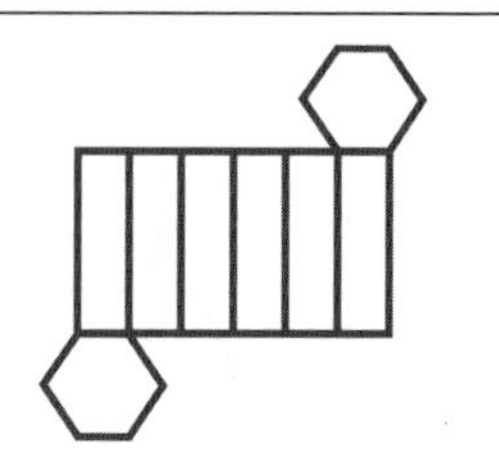

2 Draw the component 2D shapes for each 3D shape.

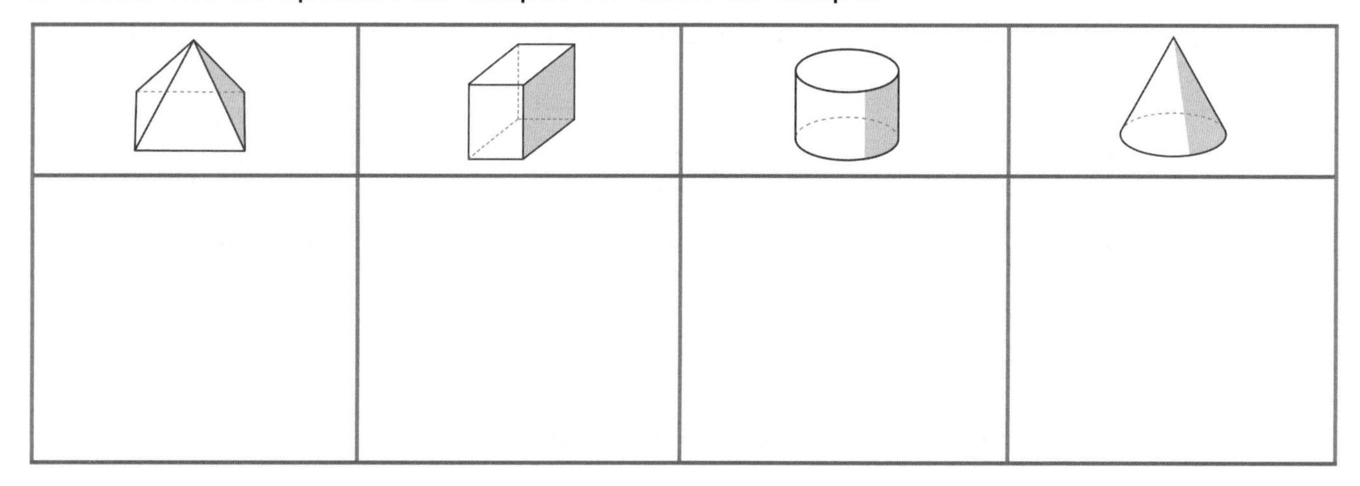

3 Draw a net for each shape.

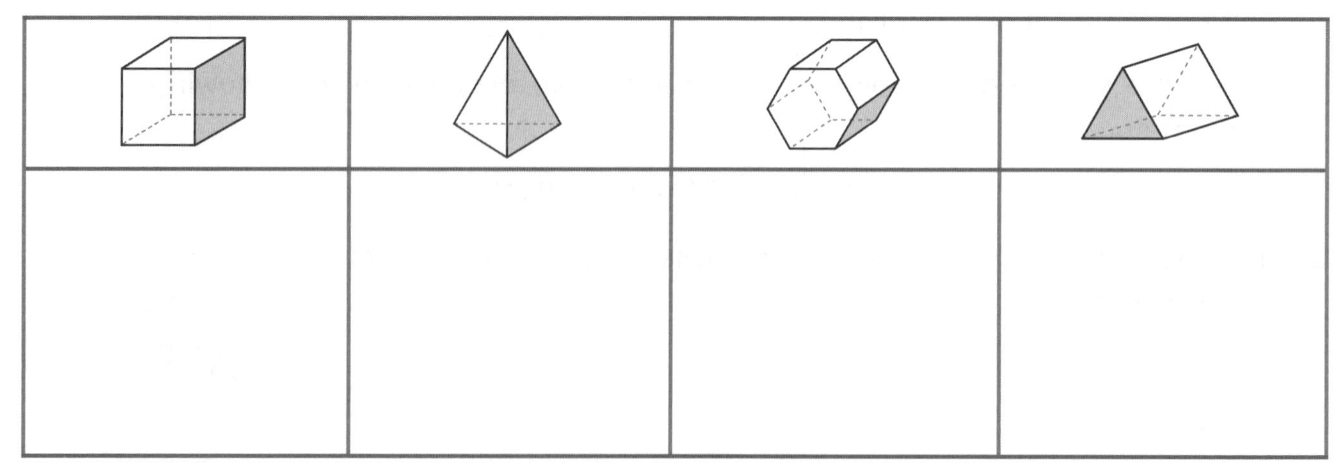

Extension: On another sheet of paper, try to draw the net of a decahedron (similar to a 10-sided dice) and a dodecahedron (like a 12-sided dice).

2D Shapes (TRB pp. 48–51)
Three-dimensional space MA3-14MG identifies three-dimensional objects, including prisms and pyramids, on the basis of their properties, and visualises, sketches and constructs them given drawings of different views
Two-dimensional space MA3-15MG manipulates, classifies and draws two-dimensional shapes, including equilateral, isosceles and scalene triangles, and describes their properties

Note: This unit focuses on two-dimesnional shapes, including those that make up the faces of three-dimensional objects, before stude move on to the specific study of prisms and pyramids in Unit 9.

DATE:

STUDENT ASSESSMENT

1 Draw:

a a trapezium **b** a right-angled triangle **c** an octagon

2 Look at each polygon.

A 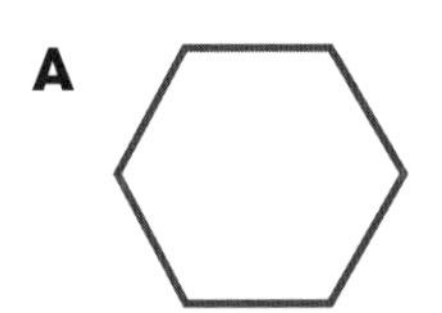 **B**  **C** 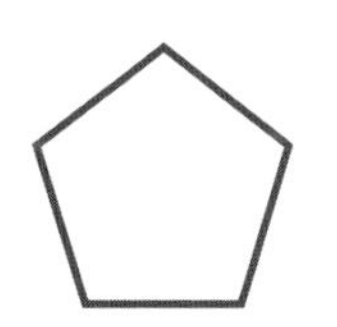 **D**  **E** 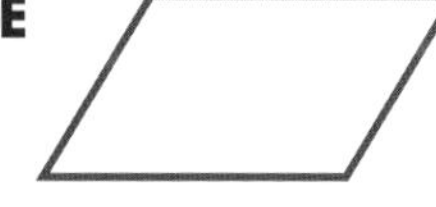

a Which shape has 5 diagonals? ________________

b Which shapes have the same number of diagonals? ________________

c Which shape has 0 diagonals? ________________

d How many diagonals does a regular hexagon have? ________________

3 Look at the labels on the circle. Which letter is pointing to:

a the circumference? ________________

b a quadrant? ________________

c the diameter? ________________

d the radius? ________________

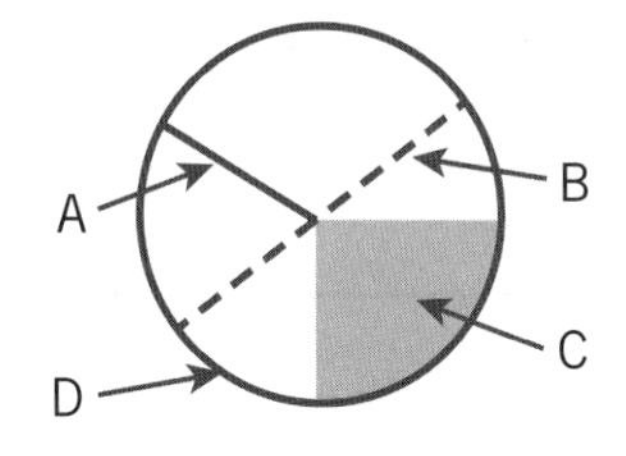

4 Draw a net for a cube.

Prism Investigation

DATE:

Investigate the historical use of prisms in building design around the world; for example, in China, Japan and Indonesia. Use the internet or a library to do your research.

Use the space below to keep note of findings from your research. There is also room at the bottom for you to draw pictures to help with your report.

Pyramids

DATE:

1 Name each 3D object.

a
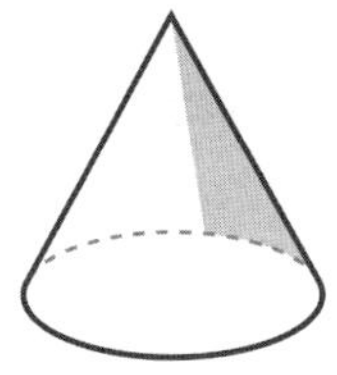

b
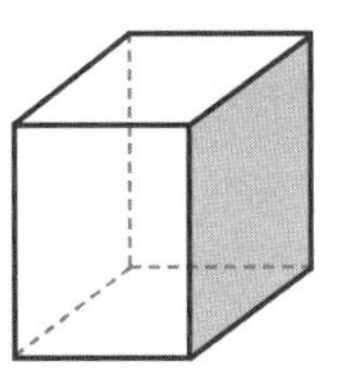

c
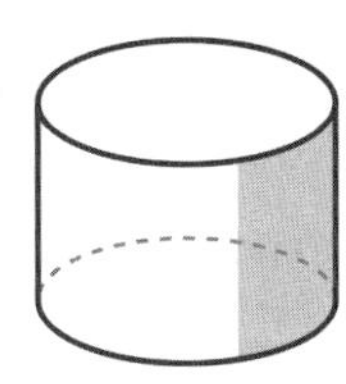

d
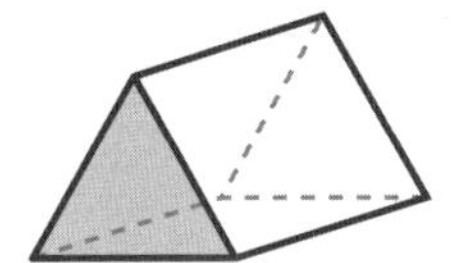

e
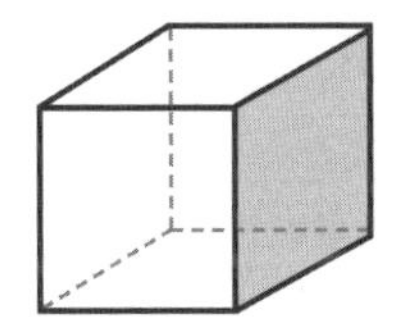

f
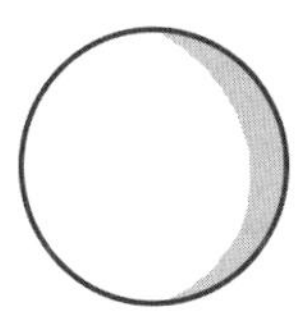

2 Identify each type of pyramid.

a
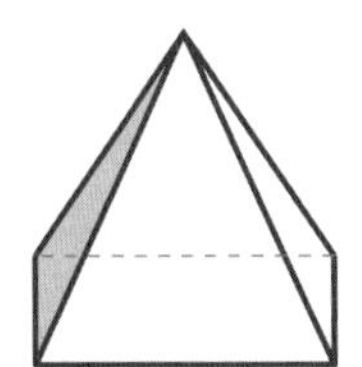

b
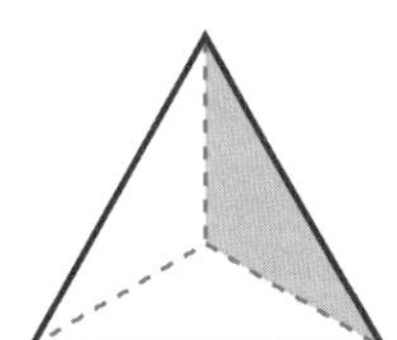

c
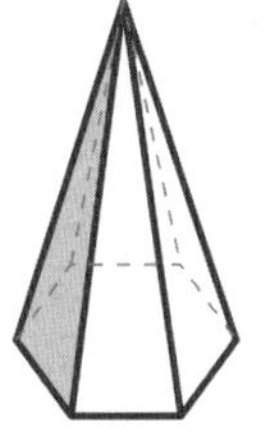

3 Draw a pyramid twice the size of the pyramid below.

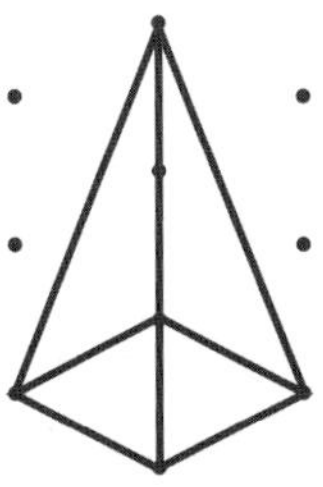

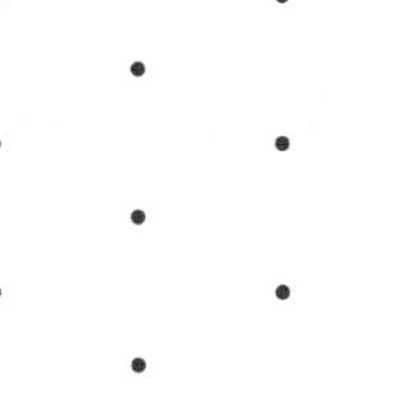

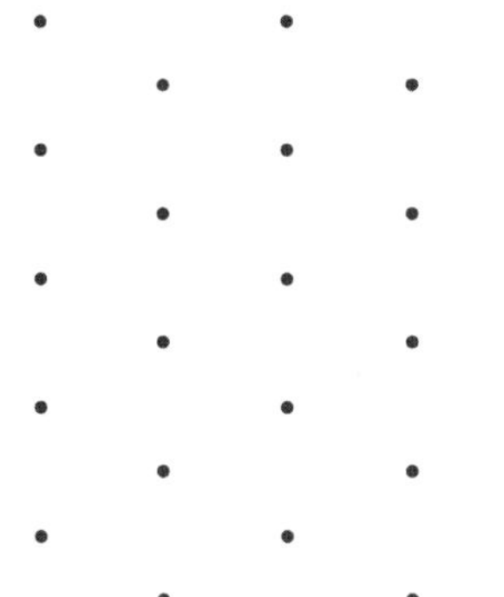

Report

DATE:

Use the questions below to assist you in making a report about your prism or pyramid from Lesson Plan 3, Independent Tasks, Task 1, "The Largest Prism or Pyramid" (Teacher's Resource Book).

My group decided to collect data on ______________________________

____

____

The way we decided to do this was ______________________________

____

____

The strength of this was ______________________________

____

____

The weakness of this was ______________________________

____

____

Our data was valid because ______________________________

____

____

____

What I learnt from collecting the data was ______________________________

____

____

____

If I did this task again, I would like to learn more about ______________________________

____

____

____

DATE:

STUDENT ASSESSMENT

1 Name each 3D object.

a 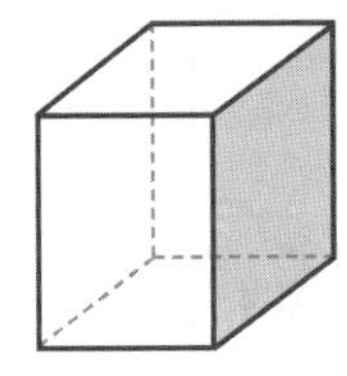

b 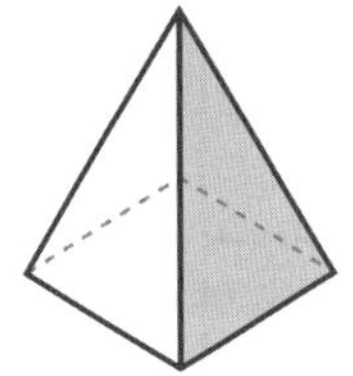

c 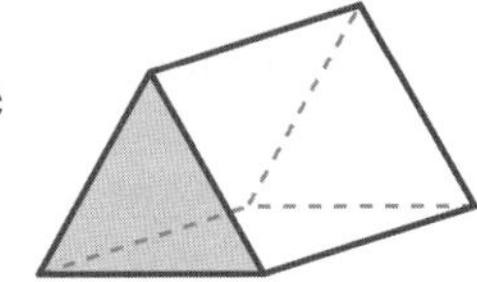

______________ ______________ ______________

2 Draw:

a a rectangular pyramid

b a hexagonal prism

3 Draw a net for a triangular-based pyramid.

4 Complete the table with the number of faces, edges and vertices of each 3D object.

	Object	Faces	Edges	Vertices
a	Cube			
b	Rectangular prism			
c	Triangular prism			
d	Square-based pyramid			
e	Triangular-based pyramid			

Coordinates

DATE:

1 Find the coordinates for the:

a giant palm tree

____

b abandoned town

____

c bear cave

____

d treasure chest

____

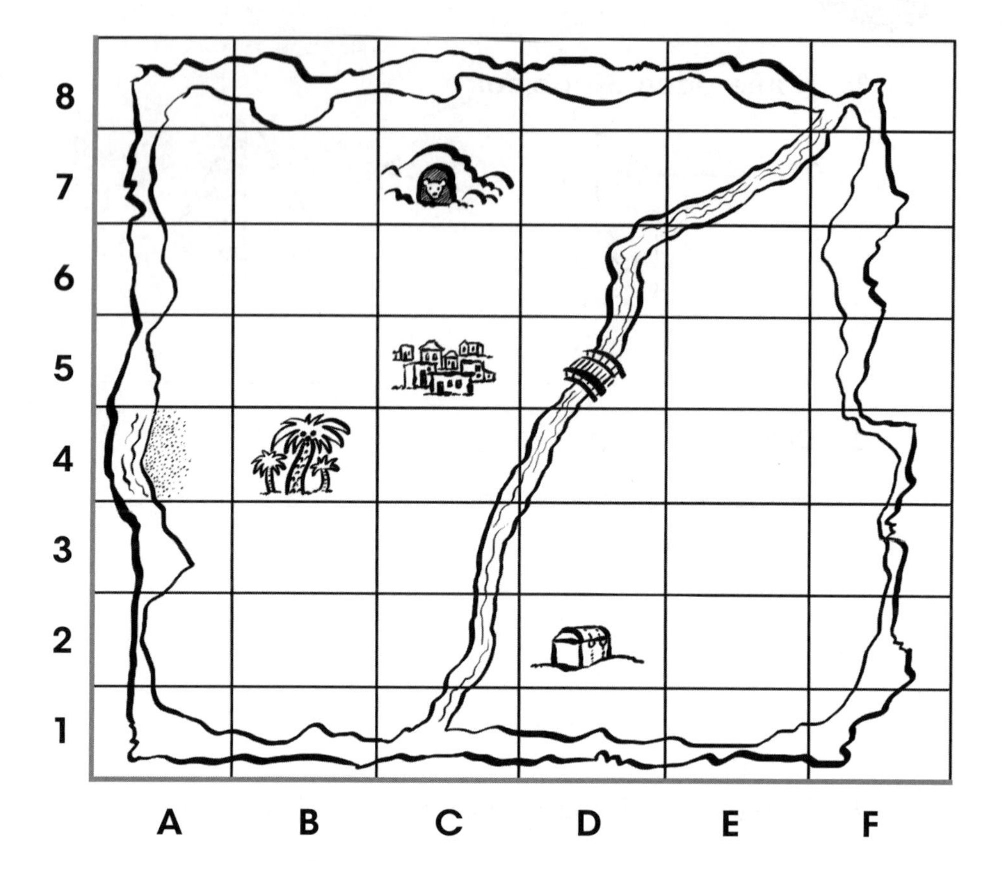

2 Place the objects at their correct coordinates.

a circle at C6

b square at E2

c diamond at A7

d star at F5

e X at E3

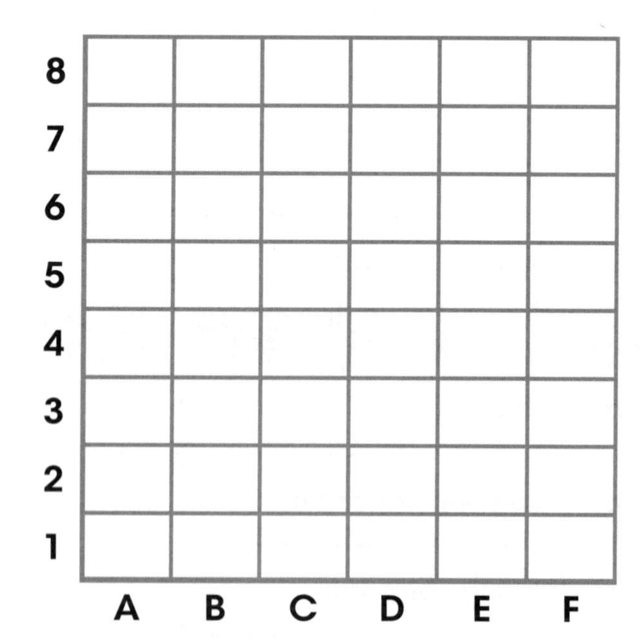

Extension: Using the map in Question 1, give directions from the beach to the treasure.

____

____

____

Mapping

DATE:

1 What can you find at the following coordinates on the map of Tasmania?

a C7 ____________ **b** H4 ____________ **c** J6 ____________

d G8 ____________ **e** E4 ____________ **f** B10 ____________

2 Where would you find the following places? Write the coordinates.

a Westbury ________ **b** Macquarie Harbour ________ **c** Longford ________

d Burnie ________ **e** Mt William Nat. Park ________ **f** Penguin ________

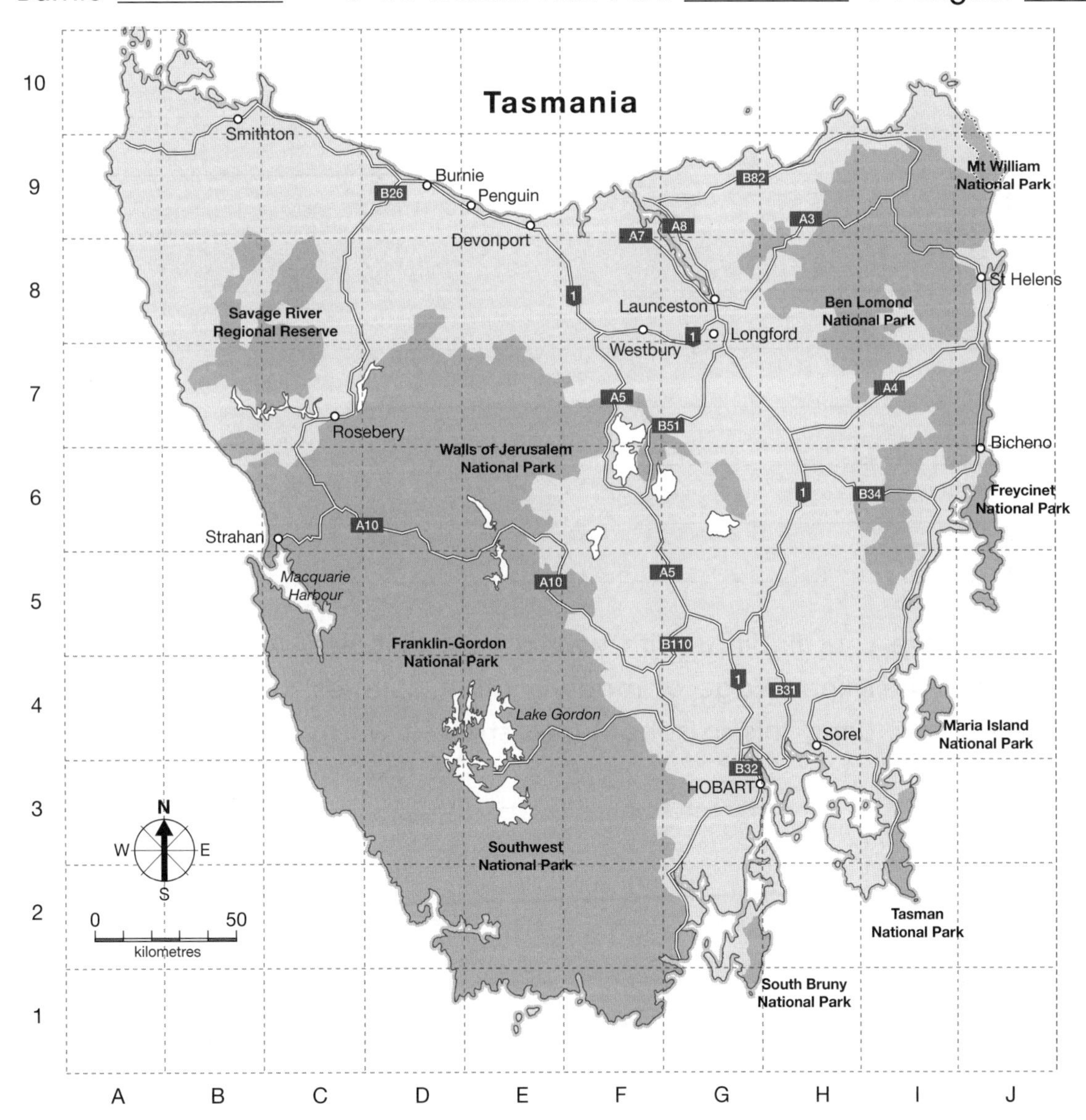

Extension: Using the roads, write directions for how to get from Hobart to Launceston.

____

____

Maps in Other Cultures

DATE:

You will need: access to a library or the internet

1 Select a culture that isn't your own. It could be from overseas or Australia.

Investigate mapping in the chosen culture. Are the maps similar to or different from the maps you're used to? How are locations indicated? Are coordinates or grid references used?

Record your findings below.

2 **Home task:** Share your findings with your parent or carer. Ask your parent or carer about their knowledge of maps in other cultures, and include that information below. Bring this to school and share it with your class.

DATE:

STUDENT ASSESSMENT

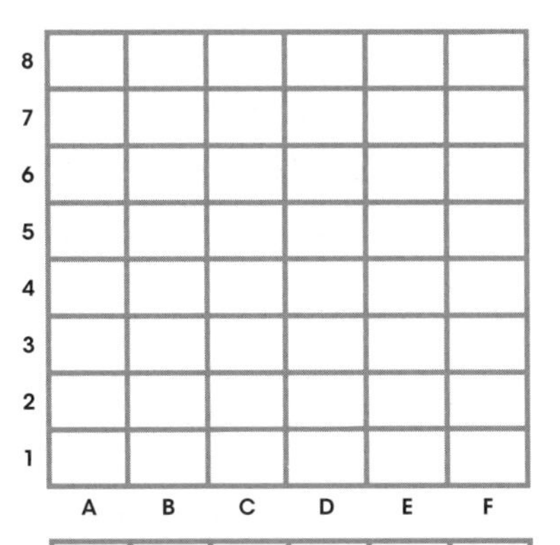

1 Draw the following symbols on the grid.

a star at A3

b square at D7

c circle at F8

2 Name the object at:

a C5 ______________________________

b E6 ______________________________

c B1 ______________________________

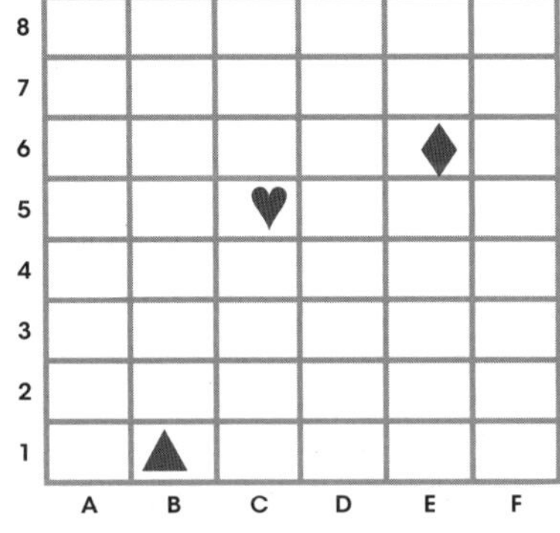

3 Why are coordinates important?
Explain with an example.

____

____

4 Using the map of Bali, explain how to get from Denpasar to Singaraja.

____

5 Using the map of Bali:

a Give two locations: ____________________ and ____________________

b Describe how to travel between these two locations.

____

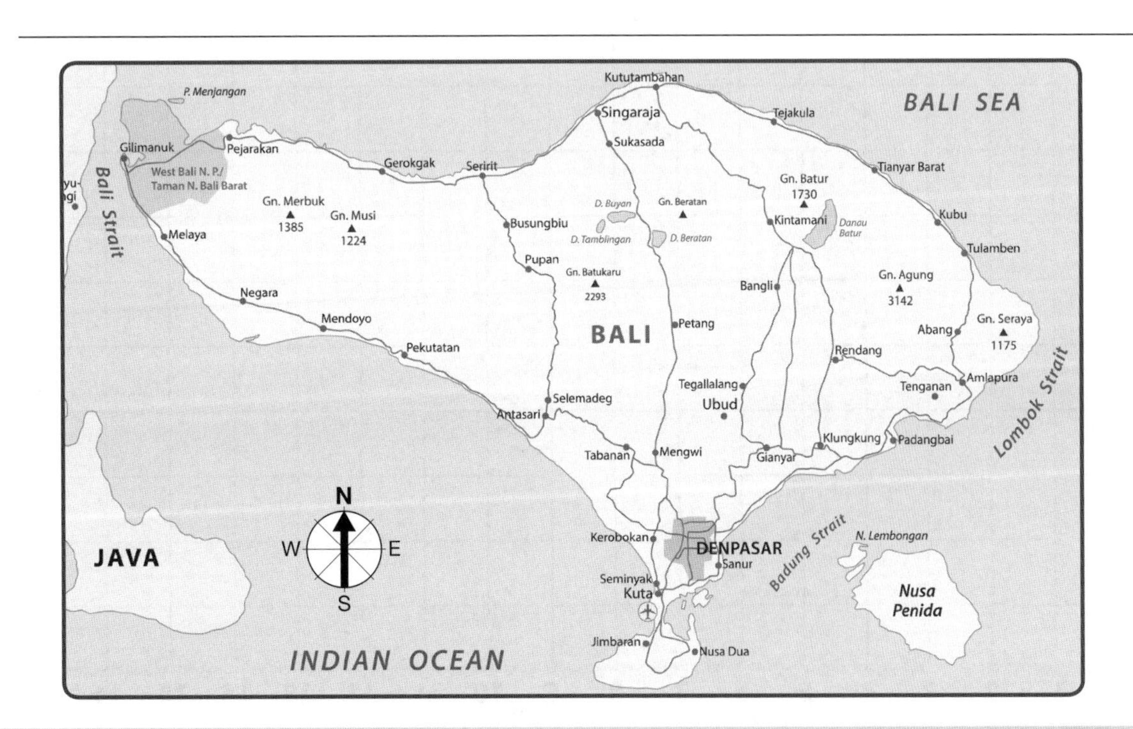

Plot to Create

DATE:

You will need: a ruler

1 Plot the following coordinates.

a Part A: (1, 5), (4, 1), (12, 1), (15, 5), (1, 5)

b Part B: (12, 5), (12, 14), (5, 6), (12, 8), (12, 5)

c Part C: (3, 5), (3, 15), (1, 13), (3, 13), (3, 5)

2 When completed, colour your picture.

3 Then add some of your own features and give their coordinates.

____

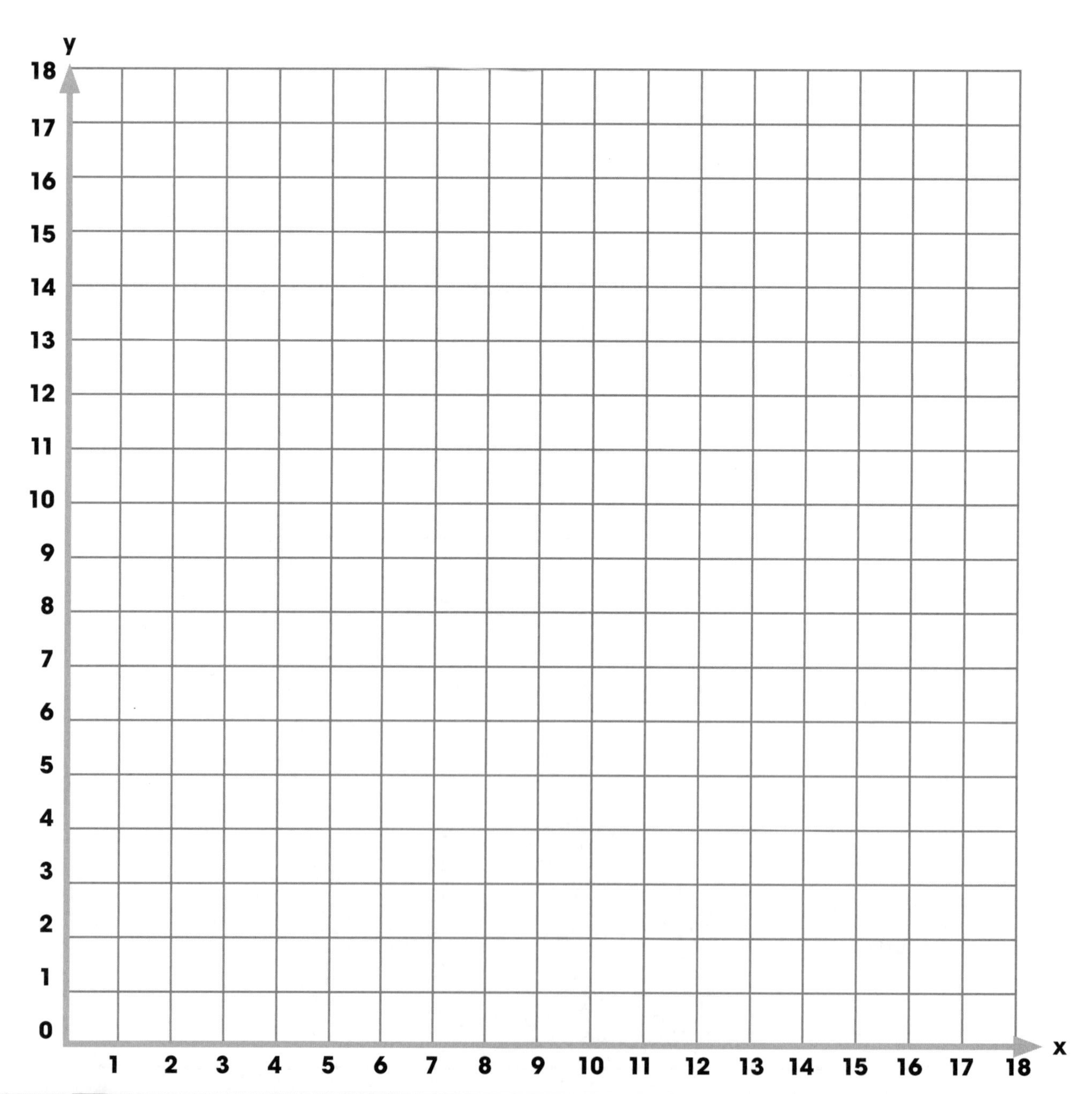

Plotting on the Cartesian Axes

DATE:

1 Plot the following coordinates. Join the points together in order.

a (3, 0) **b** (6, 4) **c** (2, 4) **d** (0, 7)

e (-2, 4) **f** (-6, 4) **g** (-3, 0) **h** (-6, -5)

i (-3, -4) **j** (0, -3) **k** (3, -4) **l** (6, -5)

m (3, 0)

2 What shape have you made? ________________

3 Add features to your picture and list the relevant coordinates.

____

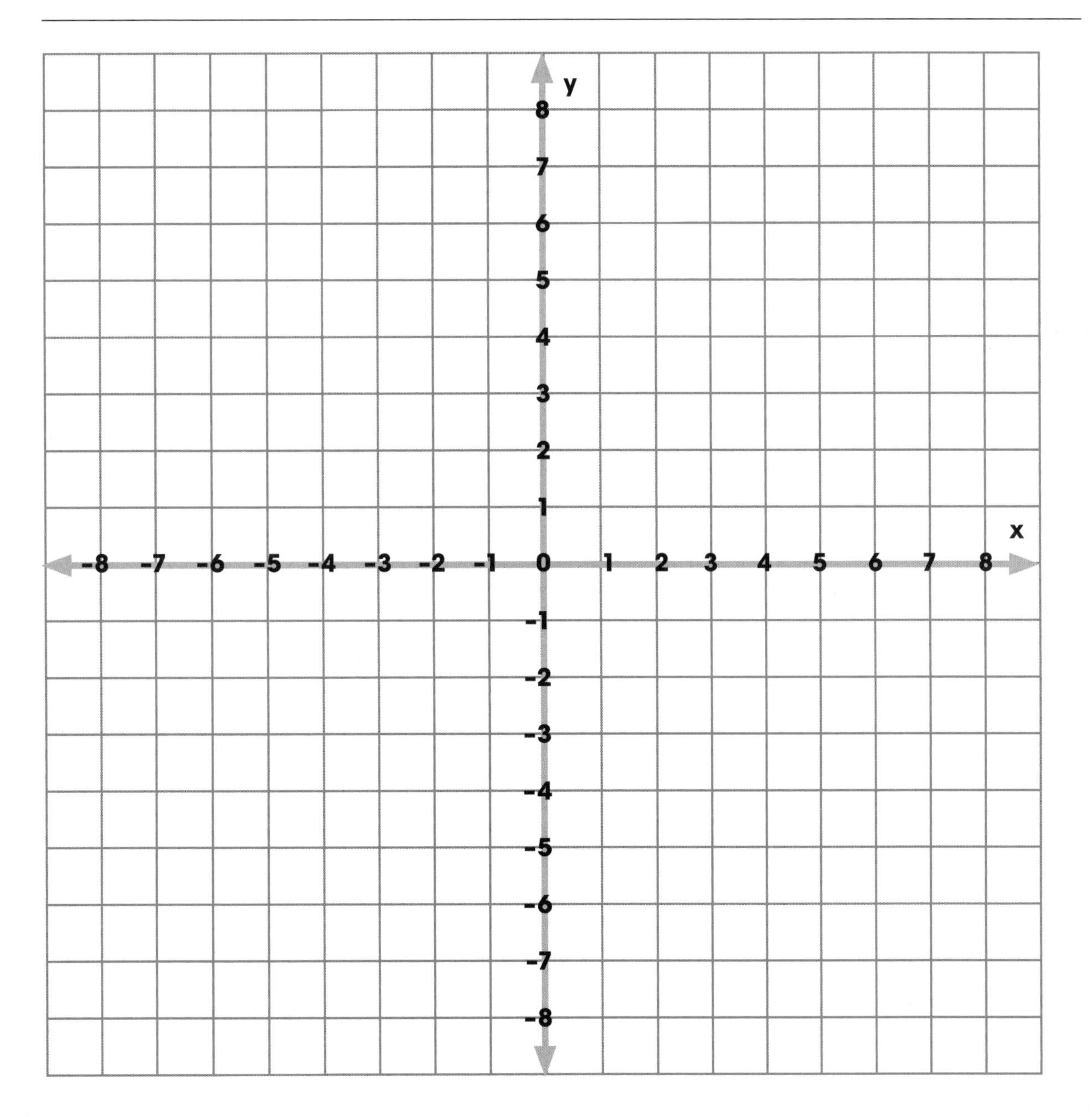

List the Coordinates

DATE:

1 List the coordinates of each of the points indicated on the image, e.g. A (-5, -6).

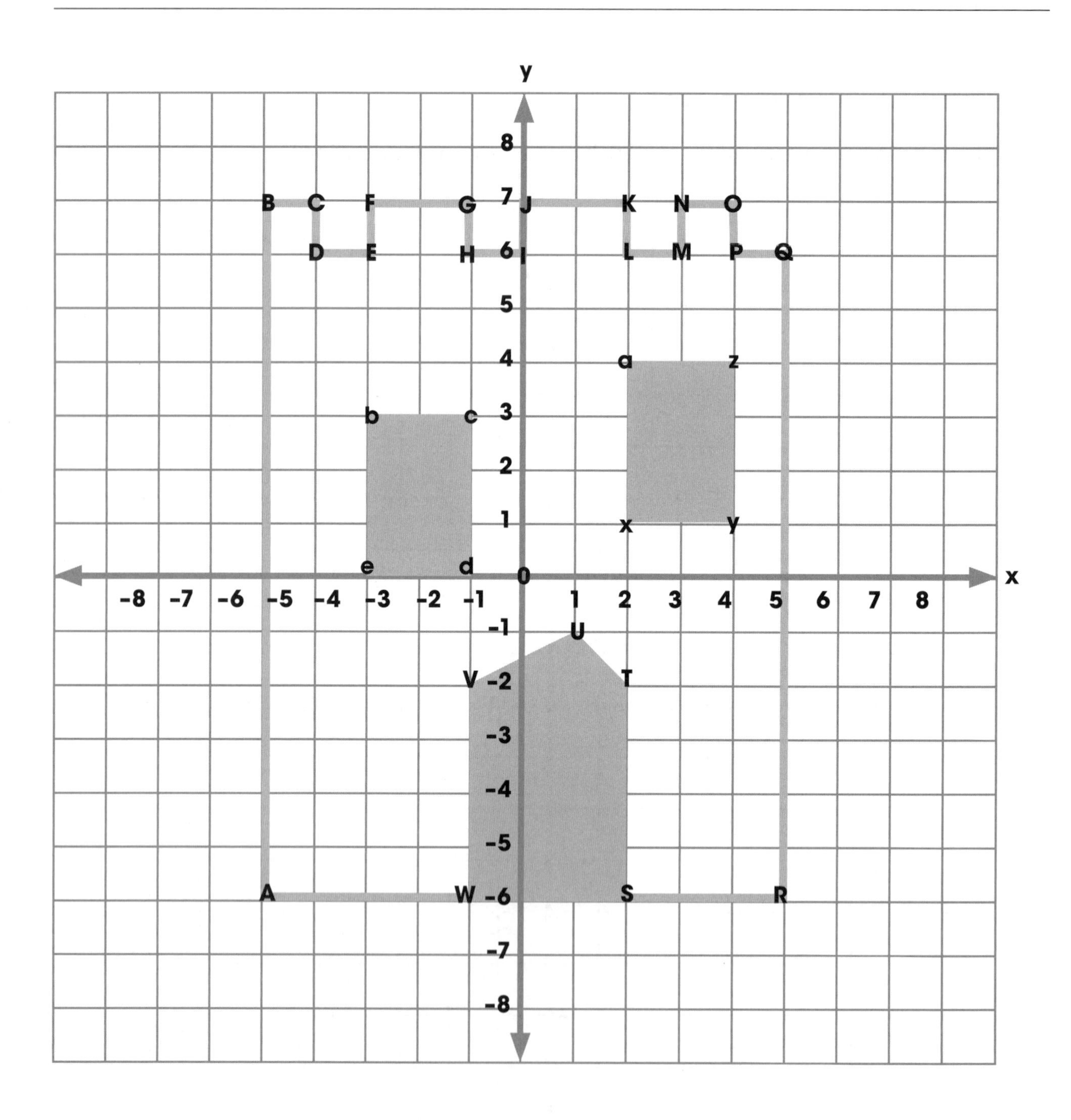

Cartesian System (TRB pp. 60–63)
Patterns and algebra MA3-8NA analyses and creates geometric and number patterns, constructs and completes number sentences, and locates points on the Cartesian plane

DATE:

STUDENT ASSESSMENT

1 List the coordinates for each of the corners of the shape.

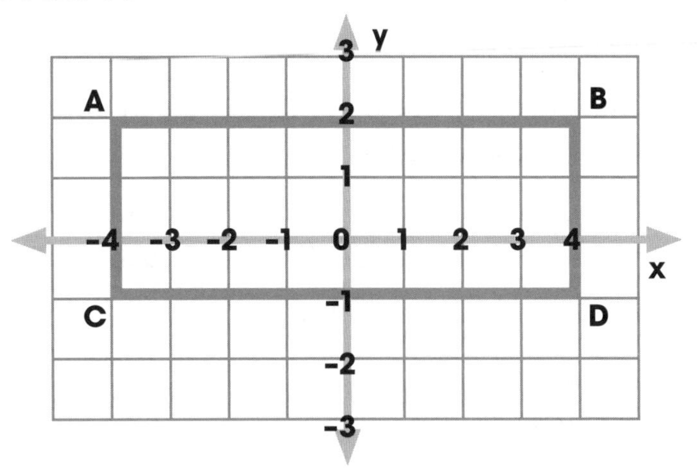

2 Plot the following points. What shape do they make?

a (0, 2) **b** (1, 1) **c** (2, 0) **d** (3, −1) **e** (4, −2)

f (−4, −2) **g** (−3, −1) **h** (−2, 0) **i** (−1, 1) **j** (0, 2)

3 Give the central coordinates for the following features:

a the school ______________ **b** the oval ______________

c the shop ______________ **d** the playground ______________

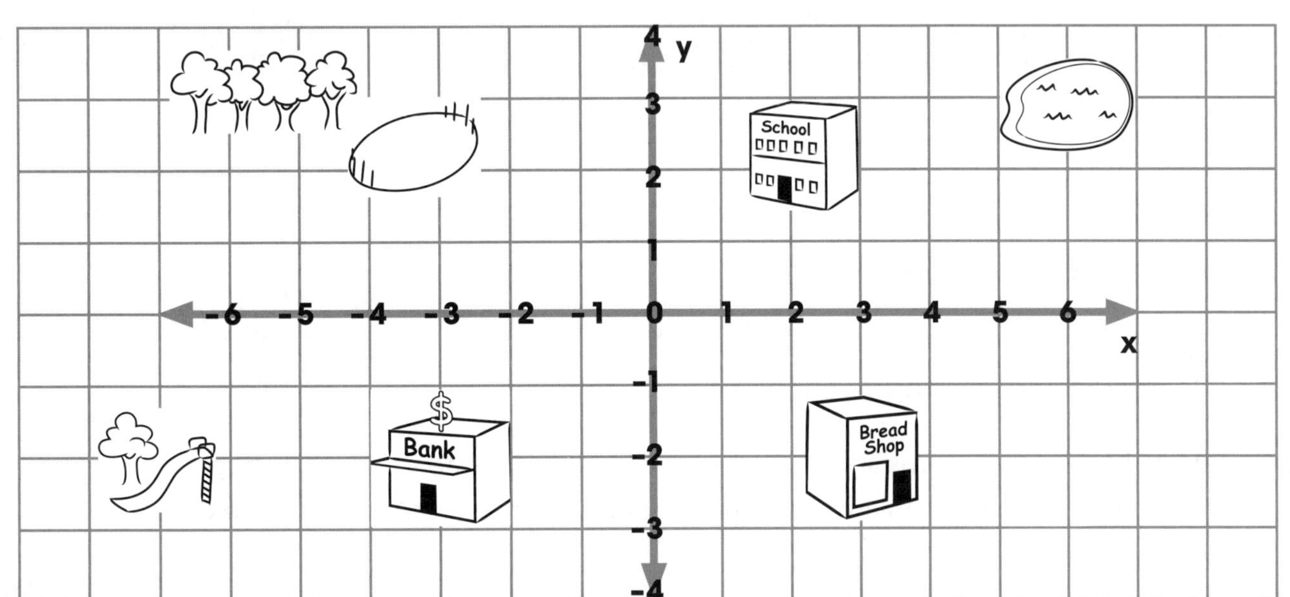

Decimal Conversions of Length

DATE:

Zoe, the builder's assistant, hasn't been very helpful with her measurements.

1 She has measured everything on the building site in different units. Fix her mistake by filling in the missing sections of the table below.

	mm	cm	m
Side wall			15.3
Front wall		825	
Roof height	2950		
Driveway length		815	
Driveway width		322.5	
Property length			24.83
Property width			15.65

2 **a** Zoe also measured a few other things, but forgot to write down what they were. Can you work out where each one belongs on the table?

garage door width   brick length   backyard length   brick width

	mm	cm	m
			0.3
		12.4	
	3480		
		855	

b Now fill in the missing sections of the table.

Extension: Measure 3 items in your classroom and create a table listing the measurements in millimetres (mm), centimetres (cm) and metres (m).

Decimal Representations of Mass

DATE:

Kia won't let you into her club unless you know the pass phrase. She has given you the answer – you just have to un-jumble it by sorting the masses from lightest to heaviest.

0.1 t	**S**
0.25 kg	**E**
$\frac{1}{2}$ t	**C**
1 kg	**E**
1 t	**R**
2 t	**Y**
10 g	**G**
75 kg	**I**
150 g	**R**
600 kg	**U**
1 200 kg	**R**
1 200 g	**N**
35 000 g	**F**
180 000 g	**H**

Pass phrase:

____ ____ ____ ____ ____   ____ ____ ____ ____

____ ____ ____ ____ ____

Extension: To become a senior leader in the club, sort these masses from lightest to heaviest.

0.05 t	5 000 kg	500 kg	5 000 g	$\frac{1}{2}$ kg

____

Decimal Conversions of Volume and Capacity

DATE:

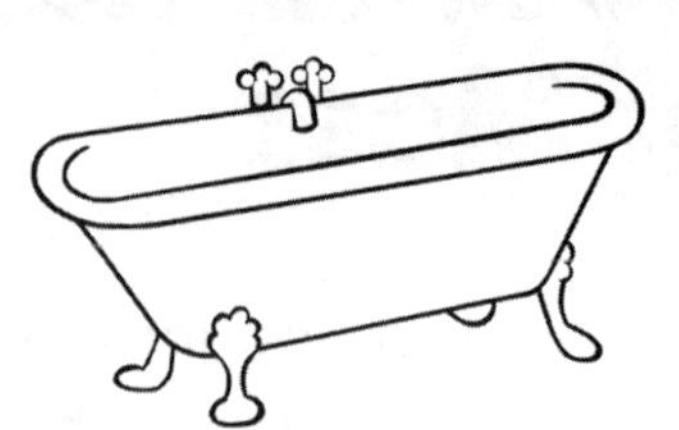

1 What unit would you use to measure the capacity of each of the following containers?

soft drink can ________ swimming pool ________ bathtub ________ cough syrup bottle ________ kettle ________

2 Fill in the missing sections of the tables below. (1 megalitre = 1 000 000 litres)

	millilitres	litres
Water bottle	600	
Large milk bottle		3
Fridge	400 000	

	litres	megalitres
Small dam		1.2
Swimming pool	2 500 000	
Small lake		3.55

3 Match the objects with their estimated capacity.

mug shoe box dam fridge test tube

4 litres 100 megalitres 400 litres 30 millilitres 300 millilitres

4 Convert the capacities into litres.

a 350 mL = ________ L **b** 700 mL = ________ L

c 1 250 mL = ________ L **d** 3 000 mL = ________ L

e 8 950 mL = ________ L **f** 15 000 mL = ________ L

Extension: Name some other things that might be measured in megalitres.

____

DATE:

STUDENT ASSESSMENT

1 Fill in the missing sections of the table.

mm	cm	m
	140	
		14.83
1 255		
	25.5	
		11.09
463		

2 Complete:

a 1 250 g = ________ kg

b ________ g = $\frac{1}{2}$ kg

c ________ g = 1.73 kg

d 42 000 g = ________ kg

3 Draw lines to match the values.

15 500 mL	0.6 L
600 mL	0.015 L
15 mL	7 L
7 000 mL	15.5 L

4 Explain how to convert between:

a centimetres and metres ____________________

b kilograms and grams ____________________

Decimal Representations of the Metric System (TRB pp. 64–67)
Length MA3-9MG selects and uses the appropriate unit and device to measure lengths and distances, calculates perimeters, and converts between units of length
Volume and capacity MA3-11MG selects and uses the appropriate unit to estimate, measure and calculate volumes and capacities, and converts between units of capacity
Mass MA3-12MG selects and uses the appropriate unit and device to measure the masses of objects, and converts between units of mass

Length Problems

DATE:

You will need: a ruler

1 Use a ruler to measure the perimeter of each shape. Make sure your answer matches the scale of measurement (cm or mm).

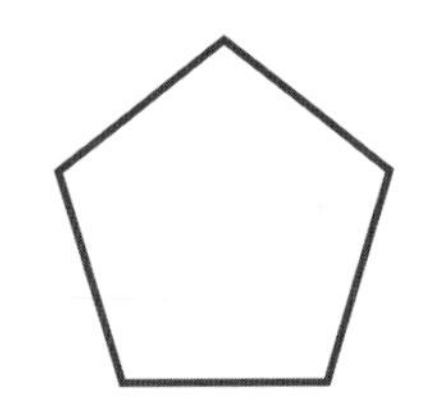

a __________ mm

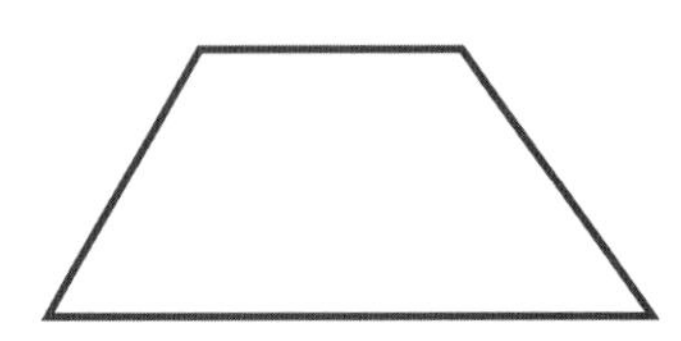

b __________ mm

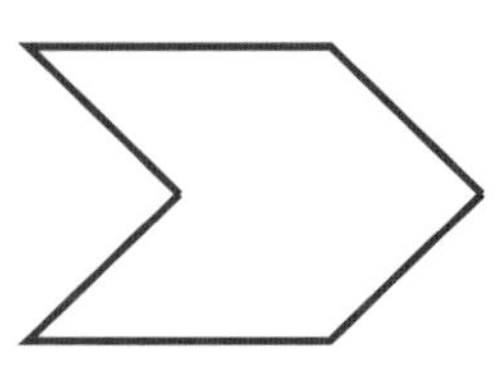

c __________ cm

d __________ mm

e __________ mm

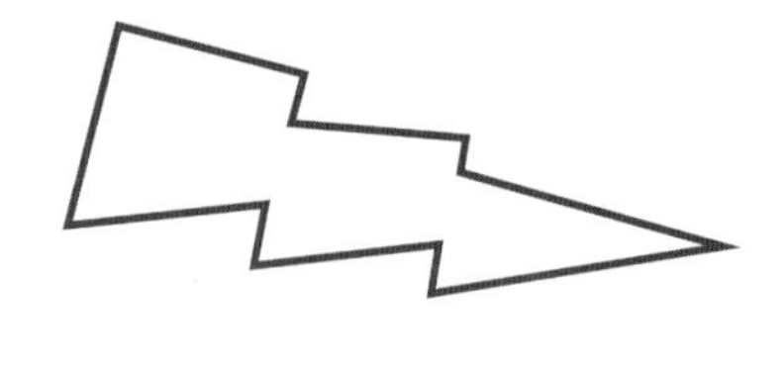

f __________ cm

g __________ mm

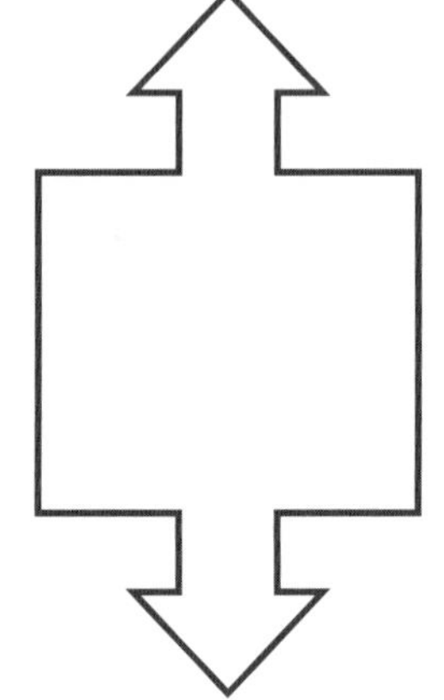

h __________ cm

2 On another sheet of paper, draw the following shapes, then work out their perimeters.

a an L-shape with the two long sides each measuring 80 mm

b a 3 cm × 4 cm rectangle attached to a 7 cm × 5 cm rectangle

c a 45 mm × 25 mm parallelogram with an equilateral triangle with sides of 45 mm on top of it

d a block arrow made up of a 7 cm × 3 cm rectangle and a 5 cm × 5 cm × 5 cm triangle

Extension: On another sheet of paper, describe in words how you will build a new shape. Then, draw it and find its perimeter.

Area Problems

DATE:

You will need: a ruler

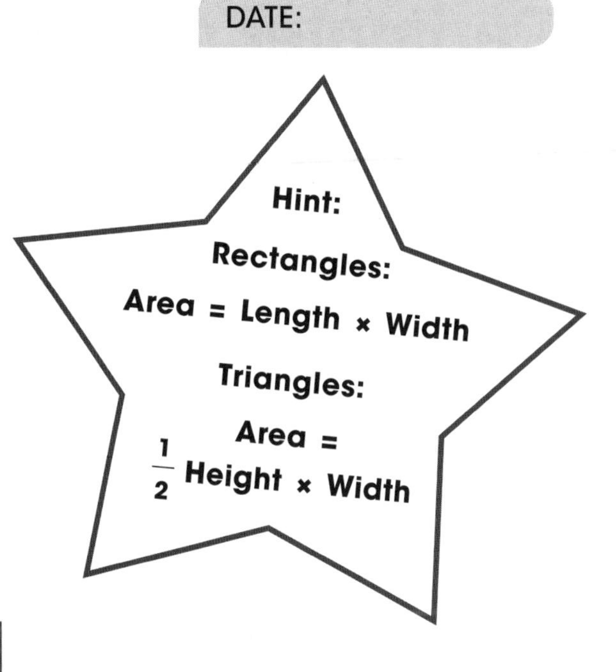

1 Using a ruler, measure the area of each shape. Make sure your answer matches the scale of measurement (cm^2 or mm^2). You may need to break the shapes up into familiar shapes.

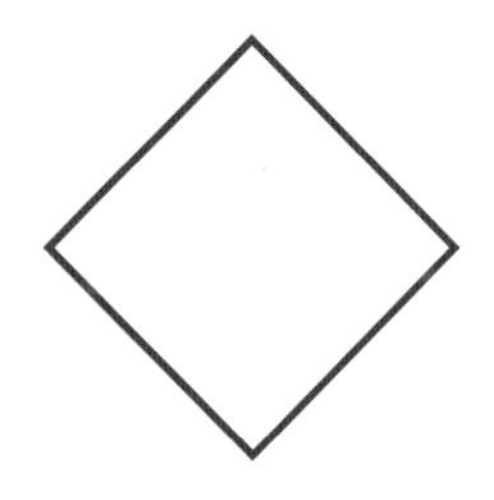

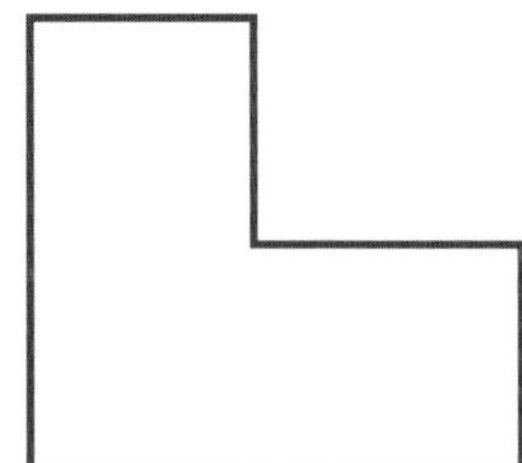

a __________ mm^2 **b** __________ mm^2

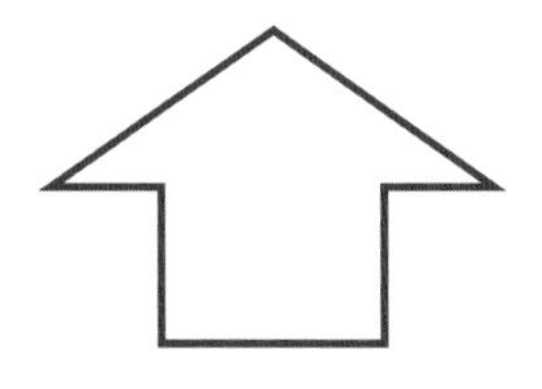

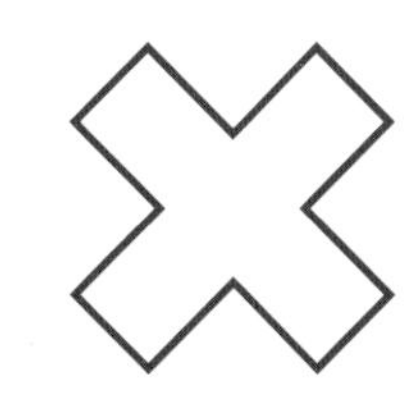

c __________ mm^2 **d** __________ mm^2

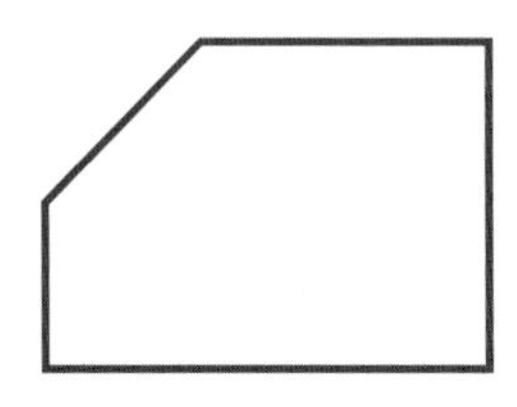

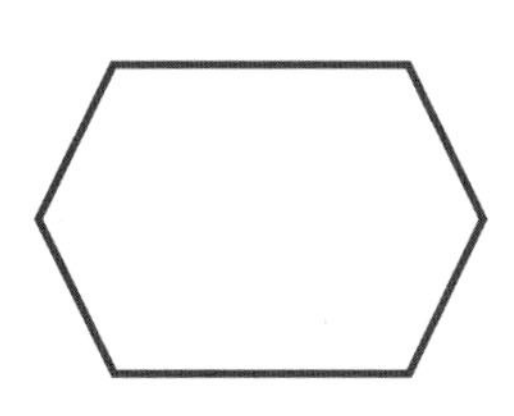

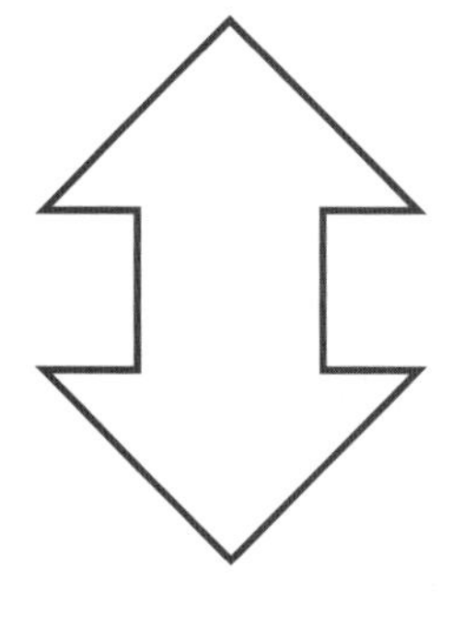

e __________ cm^2 **f** __________ cm^2 **g** __________ cm^2

2 On another sheet of paper, draw the following shapes, then work out their areas.

a a rectangle with a length of 7 cm and a total perimeter of 22 cm

b an equilateral triangle with 60 mm sides

c an 8.5 cm × 5.5 cm parallelogram

Perimeter and Area Problems

DATE:

1 Work out the area of each shape.

a 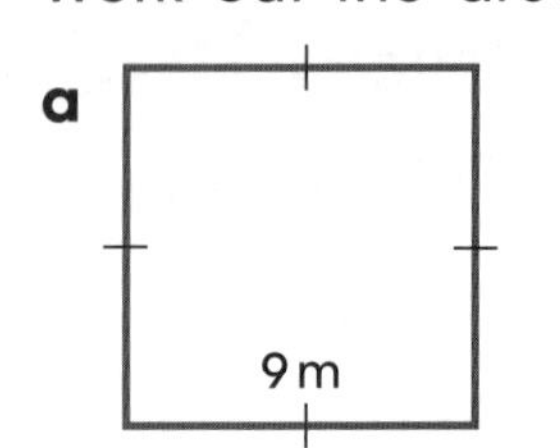

b 6 m
12 m

c 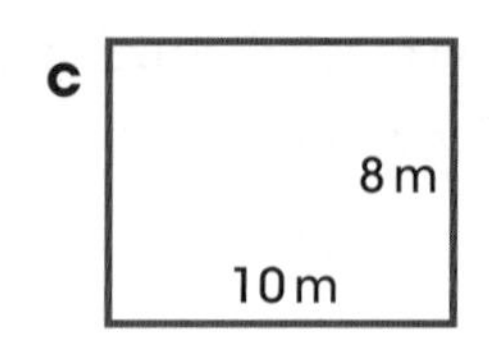

d 2 m
16 m

A = ________ A = ________ A = ________ A = ________

e What do these shapes have in common? ______________________________

2 All of the shapes below have an area of 36 m². Work out the missing side lengths.

a 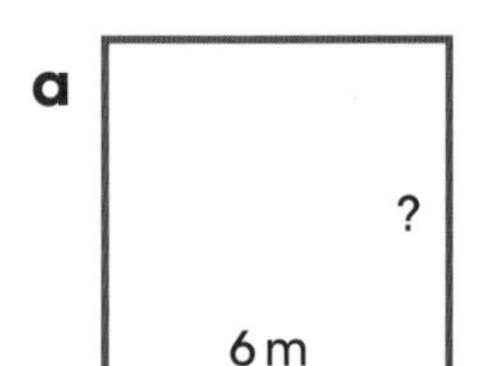

b 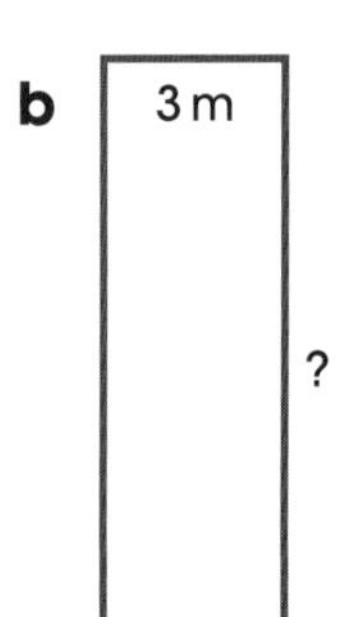

c 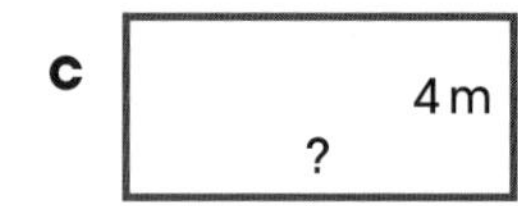

d 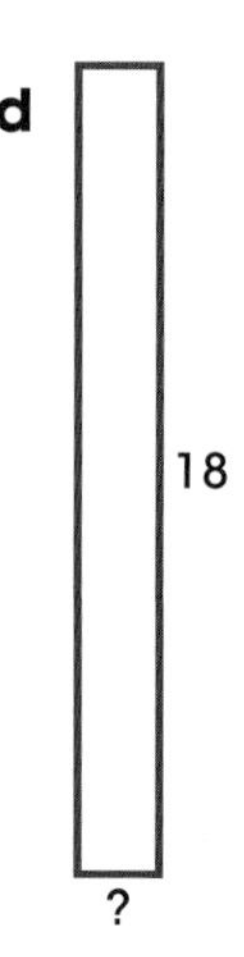

3 On another sheet of paper, use the perimeter measurements below to find squares or rectangles with the **smallest** and the **largest** areas possible. Record your findings here:

a 16 m ______________________________

b 26 m ______________________________

c 36 m ______________________________

Extension: On another sheet of paper, show four different ways you could build a fence around an area of 96 m².

Length and Area Problems (TRB pp. 68–71)
Length MA3-9MG selects and uses the appropriate unit and device to measure lengths and distances, calculates perimeters, and converts between units of length
Area MA3-10MG selects and uses the appropriate unit to calculate areas, including areas of squares, rectangles and triangles

DATE:

STUDENT ASSESSMENT

1 Find the perimeter of each shape.

a

6 cm

b

6 cm
5 cm
8 cm

c

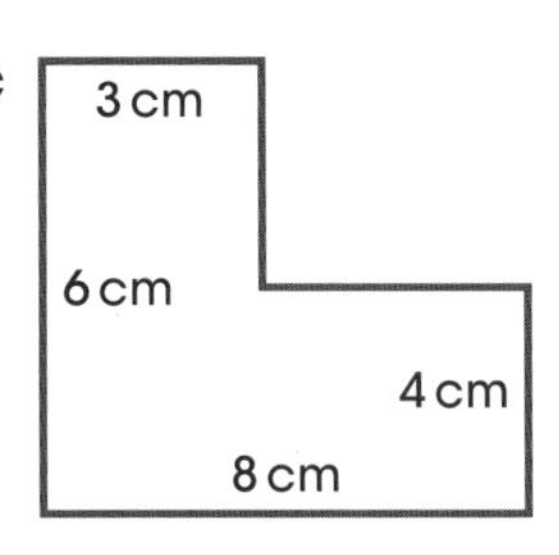

P = ____________ P = ____________ P = ____________

2 Find the area of each shape.

a

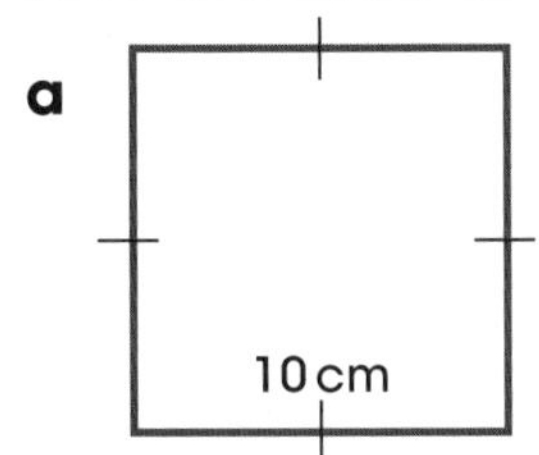

b

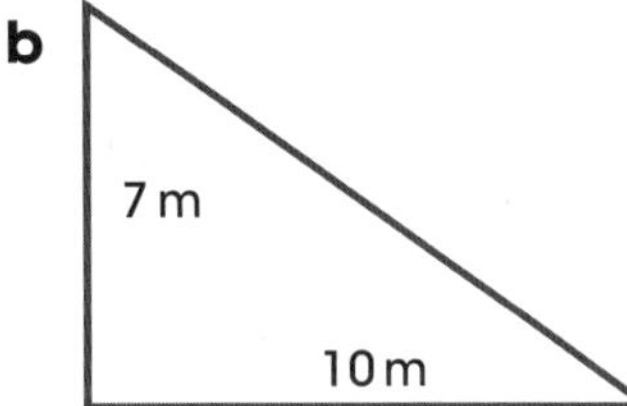

c

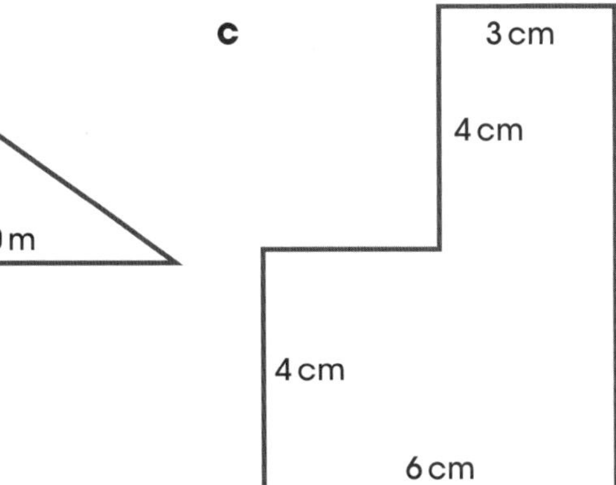

A = ____________ A = ____________ A = ____________

3 All of these shapes have an area of $24\,m^2$. Find the missing side lengths.

a

?
4 m

b

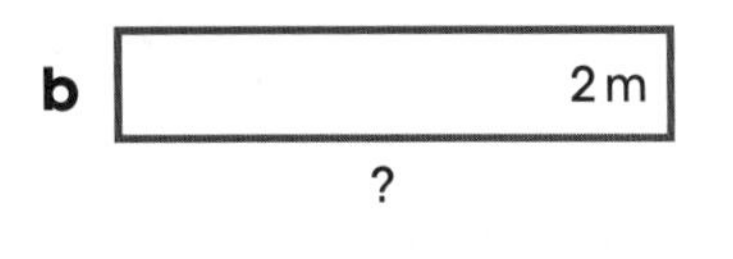

c

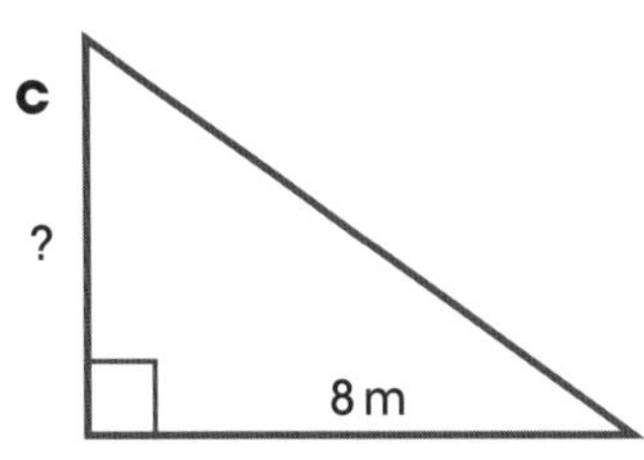

4 On another sheet of paper, draw a shape with a perimeter of:

a 50 cm **b** 28 cm

Rounding Numbers

DATE:

1 Round each value to the nearest whole number.

a 7.9 ________ **b** 3.7 ________ **c** 10.41 ________ **d** 14.81 ________

e 25.9 ________ **f** 33.59 ________ **g** 48.18 ________ **h** 93.61 ________

i 243.01 ________ **j** 120.87 ________ **k** 399.75 ________ **l** 563.19 ________

2 Round each value to **one** decimal place.

a 4.16 ________ **b** 14.73 ________ **c** 6.83 ________ **d** 12.11 ________

e 31.92 ________ **f** 44.44 ________ **g** 54.99 ________ **h** 88.07 ________

i 167.035 ________ **j** 249.166 ________ **k** 831.709 ________ **l** 738.463 ________

3 Using your rounding skills, estimate each answer as a whole number.

a 13.8 + 16.3 = **b** 11.21 + 8.8 =

c 28.67 + 11.3 = **d** 38.93 + 19.12 =

e 52.88 + 17.2 = **f** 68.11 + 31.91 =

g 18.33 - 8.75 = **h** 22.43 - 18.55 =

I 31.02 - 10.89 = **j** 48.64 - 12.15 =

4 Estimate the total of each shopping list.

a

apples	$4.99
pears	$1.48
oranges	$4.99
strawberries	$4.99
Estimate =	$________

b

beans	$4.98
carrots	$3.29
potatoes	$4.05
pumpkins	$2.65
Estimate =	$________

Extension: Do these estimations look right? Why / why not?

a 43.18 + 14.73 = 57 ________________________________

b 38.79 - 15.22 = 25 ________________________________

Adding Decimals

DATE:

1 Use rounding to estimate an answer to each problem.

a 14.8 + 11.2 =

b 19.61 + 13.25 =

c 24.12 + 18.9 =

d 38.02 + 14.88 =

2 Work out the answer to each problem.

a 38.2 + 15.4

b 46.3 + 24.6

c 52.5 + 33.7

d 268.9 + 139.5

e 14.34 + 11.45

f 38.90 + 13.44

g 58.65 + 45.14

h 78.27 + 43.47

i 17.254 + 12.481

j 24.177 + 21.439

k 58.125 + 45.931

l 73.667 + 46.688

3 Jen saved $13.85 worth of silver coins. Her mum found another $8.55 worth for her. How much did Jen have in total?

4 Yang bought a pineapple for $3.89 and a peach for $1.54. How much did he spend?

5 Quentin scored 84.25 on his first dive and 81.75 on his second dive. What was his total?

Extension: On another sheet of paper, write and then solve your own addition equation, using two double-digit numbers with 3 decimal places. You can't repeat the same digit anywhere in the question.

Subtracting Decimals

DATE:

1 Use rounding to estimate an answer to each problem.

a 16.3 - 12.1 =

b 23.18 - 12.23 =

c 31.47 - 18.94 =

d 47.5 - 38.62 =

2 Work out the answer to each problem.

a 42.3 - 25.1

b 38.7 - 22.3

c 64.8 - 35.7

d 143.2 - 127.5

e 34.53 - 26.12

f 46.82 - 27.74

g 73.21 - 46.74

h 87.72 - 31.96

i 19.187 - 14.438

j 15.00 - 7.05

k 50.00 - 23.26

l 100.00 - 57.39

3 Samira has $43.55 in her bank account. If she bought a new pair of shoes for $29.95, how much would she have left?

4 Gareth was given $50.00 for his birthday. He immediately bought a CD for $23.95. How much birthday money was left?

5 Anton had a 30 m ball of string. He cut a length of 8.75 m. How much string did he have left?

Extension: If you use a $20 note to buy a $9.55 pencil sharpener, a $2.15 eraser and a $5.95 pen, how much will you have left?

STUDENT ASSESSMENT

1 Round each value to the nearest whole number.

a 28.38 __________ **b** 57.86 __________ **c** 101.2 __________

2 Round each value to **one** decimal place.

a 88.43 __________ **b** 101.78 __________ **c** 26.04 __________

3 Complete:

a 12.7 + 3.9 =

b 17.62 + 14.73 =

c 9.36 + 17.43

d 18.36 + 7.92

e 73.684 + 10.487

4 Complete the problems.

a 21.3 - 4.7 =

b 42.6 - 21.83 =

c 9.36 - 2.21

d 5.302 - 1.731

e 95.426 - 17.218

5 Complete the problems.

a Joshua has three lengths of wood: 2.81 m, 3.46 m and 1.76 m. What is the total length of wood Joshua has?

b Ari had $5.00 pocket money. He bought some trading cards for $3.75. How much money did he have left?

c Julie has 2 L of water in a jug. If she pours 125 mL into one container and 370 mL into another, how much water will be left in the jug?

Multiplying Decimals

DATE:

1 Find the solutions to these multiplication equations featuring tenths.

a 0.5×8 **b** 0.2×6 **c** 1.4×3 **d** 5.5×4 **e** 4.3×9 **f** 6.7×5

g 7.4×7 **h** 0.8×8 **i** 5.1×9 **j** 6.9×3 **k** 9.7×6 **l** 5.3×7

2 Find the solution to each problem.

a Hayley has six 1.25 L bottles of green tea. How much green tea does she have in total?

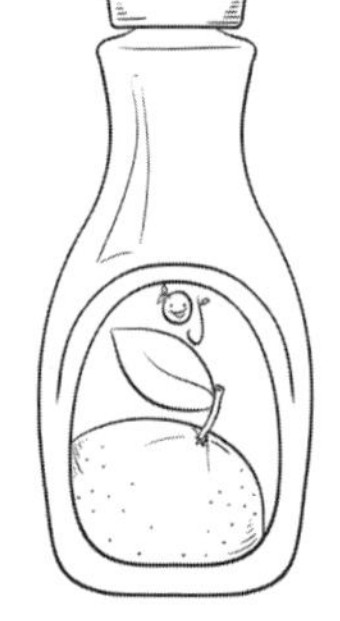

b Kyle traded in 9 orange juice bottles for recycling for $0.35 each. How much did he earn?

c Farid started with $6.37 in his account. He finished with five times that amount. How much did he end up with?

Extension:

a 4.125×3 **b** 6.333×5 **c** 0.714×9 **d** 6.305×7 **e** 9.258×6 **f** 7.044×8

Melody's Multiples

DATE:

Melody has been booked to play at a centipede's party. However, the centipede has requested that Melody doesn't play one particular song.

Solve the multiplication problems, then sort them from **smallest** to **largest** to find the name of the song.

P	Y	E	E	O
0.2 × 10	0.98 × 10	0.012 × 10	0.104 × 10	0.29 × 10

H	T	K	O	Y
0.006 × 10	0.0003 × 100	0.036 × 100	0.0065 × 100	0.014 × 100

E	H	K
0.0037 × 1000	0.0018 × 100	0.0098 × 100

Answer:

___ ___ ___ ___ ___ ___ ___ ___

___ ___ ___ ___ ___

Why doesn't the centipede want this song?

____

____

Dividing Decimals

DATE:

1 Find the solutions to these division equations.

a $4\overline{)3.2}$ **b** $6\overline{)4.8}$ **c** $3\overline{)2.7}$ **d** $5\overline{)5.5}$ **e** $8\overline{)6.4}$

f $7\overline{)4.9}$ **g** $6\overline{)3.6}$ **h** $4\overline{)5.2}$ **i** $9\overline{)10.8}$ **j** $7\overline{)8.4}$

2 Find the solutions to these division equations.

a $10\overline{)6.8}$ **b** $10\overline{)7.4}$ **c** $10\overline{)3.2}$ **d** $20\overline{)4.6}$ **e** $50\overline{)5.5}$

f Do you notice a pattern in this set of equations?

____

3 Find the solutions to these equations.

a $3.8 \div 4 =$ **b** $14.6 \div 10 =$ **c** $9.5 \div 2 =$

d $9.9 \div 5 =$ **e** $8.4 \div 5 =$ **f** $7.3 \div 8 =$

4 Find:

a 5.2 divided by 8 ________

b 9.3 divided by 6 ________

c 10.5 divided by 6 ________

Extension: Find the solutions to these equations.

a $16.8 \div 4 =$ **b** $28.9 \div 20 =$ **c** $15.8 \div 5 =$

d $46.3 \div 2 =$ **e** $37.45 \div 5 =$ **f** $75.6 \div 8 =$

DATE:

STUDENT ASSESSMENT

1 Complete:

a 5.5 × 6 = ____

b 9.2 × 3 = ____

c 8.1 × 7 = ____

d 1.91 × 5 = ____

e 4.26 × 4 = ____

2 Find:

a 0.3 × 10 =

b 0.0042 × 1 000 =

c 1.36 × 100 =

d 4.21 × 10 =

e 9.032 × 1 000 =

f 0.423 × 100 =

3 Complete:

a 9) 54

b 8) 48

c 5) 60

4 Find:

a 14.8 ÷ 10 =

b 12.01 ÷ 10 =

c 9.36 ÷ 100 =

d 0.367 ÷ 100 =

5 Find the solutions.

a Rosa has 6 bags of peas weighing 0.6 kg each. What is the total weight of the peas?

b Ling had 93.6 m of fabric. He divided it evenly between 3 customers. What length of fabric did each customer receive?

Find the Volume

DATE:

You will need: a container, a measuring cup, some substances to put in the container (e.g. sand, water, sugar), scales

1 Draw your container here.

2 **a** What is the volume of the container? ___________

b Explain how you found this.

____

____

3 **a** Did you work out the volume more than once? ___________

b Why? ____

c Was it different the second time? ___________

4 **a** What is the mass of the container when it is full? ___________

b What was in the container when you measured its mass? ___________

c Does the container's mass change with a different substance inside?

____

Capacity

DATE:

1 Find the capacity of each jug.

a

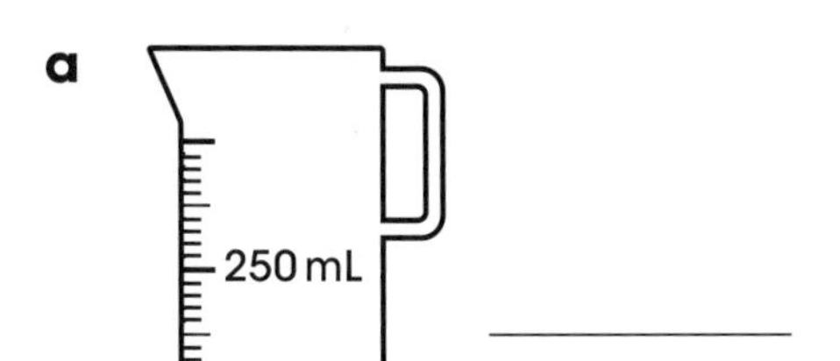

____

b

1 L

____

c

300 mL

____

d

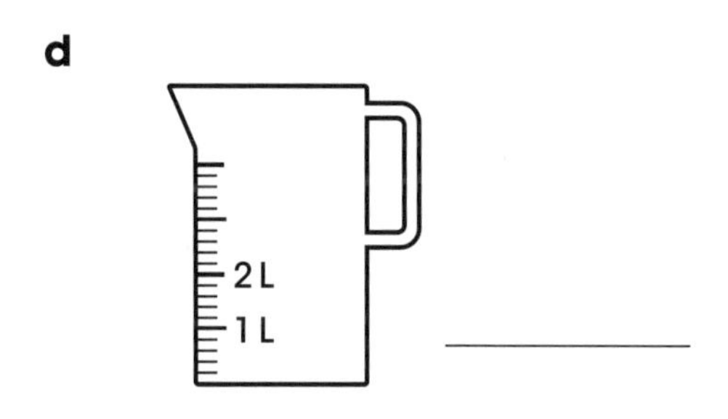

____

2 Use decimal notation to write the amounts in **litres**.

a 1 500 mL __________ L **b** 3 650 mL __________ L **c** 750 mL __________ L

d 1 480 mL __________ L **e** 1 250 mL __________ L **f** 80 mL __________ L

3 How many millilitres are in the following?

a 2.4 L __________ mL **b** 0.9 L __________ mL **c** 8.6 L __________ mL

d 12.3 L __________ mL **e** 32.1 L __________ mL **f** 100 L __________ mL

4 What is the capacity (in litres) of a jug that can exactly fill:

a 6 × 250 mL glasses __________ L **b** 10 × 300 mL cups __________ L

c 4 × 375 mL mugs __________ L **d** 3 × 800 mL bottles __________ L

5 If one litre of water is equal to one kilogram, what is the mass of the following containers when they are full of water?

a 5 L bucket __________ kg **b** 300 mL mug __________ kg

c 80 L tank __________ kg **d** 3 750 mL jug __________ kg

Extension: On another sheet of paper, create a list of common containers used for drinking. List their capacity and their weight when full of water.

Capacity and Volume

DATE:

1 Complete the following table.

	Length (m)	Breadth (m)	Height (m)	Volume (m^3)
a	3	2	1	
b	6	5	3	
c	12	5	5	
d	20	20	10	
e	12	8	6	
f	50	25	2	

2 How many millilitres of water would be displaced if you put these objects in a bucket full of water?

a 10 cm^3 eraser ___________ **b** 54 cm^3 deck of cards ___________

c 450 cm^3 cheese block ___________ **d** 200 cm^3 carton of milk ___________

3 How many centicubes must have been put in a tank if the following amounts of water were displaced?

a 12 mL ___________ **b** 73 mL ___________ **c** 125 mL ___________

d 380 mL ___________ **e** 2 L ___________ **f** 9.1 L ___________

4 Complete the table.

	Length (cm)	Breadth (cm)	Height (cm)	Volume (cm^3)	Capacity (mL)
a	6	5	1		
b	9	7	2		
c	10	5		150	
d	15		4		180
e		8	6	960	
f	12		7		1680

Extension: If 1 cm^3 = 1 mL, does 1 m^3 = 1 L? Explain your answer.

___

___

DATE:

STUDENT ASSESSMENT

1 **a** What is the difference between **capacity** and **volume**?

____

____

b What is the relationship between **capacity** and **volume**?

____

____

2 Complete the table.

	Length (cm)	Width (cm)	Height (cm)	Volume (cm^3)
a	8	5	1	
b	2	4	3	
c	4	2	5	
d	9	10	20	

3 Express each of the following in litres.

a 700 mL __________ L **b** 1 430 mL __________ L

c 50 mL __________ L **d** 12 496 mL __________ L

4 Express each of the following in millilitres.

a 9.30 L __________ mL **b** 0.4 L __________ mL

c 1.76 L __________ mL **d** 30 L __________ mL

5 How many centicubes must have been put in a tank if the following amounts of water were displaced?

a 90 mL __________ **b** 4.1 L __________ **c** 3.25 L __________

6 Find the volume of each prism.

a

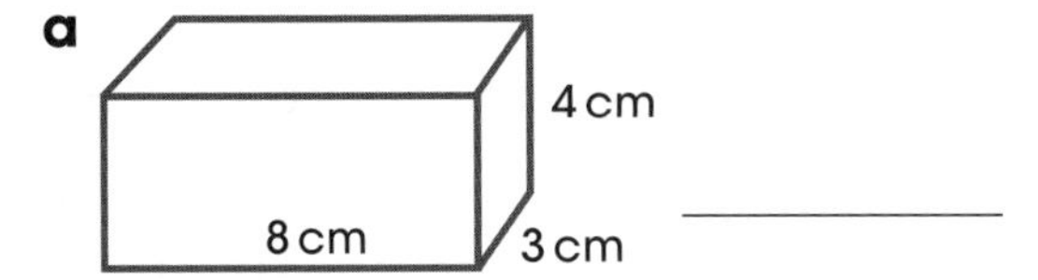

____

b

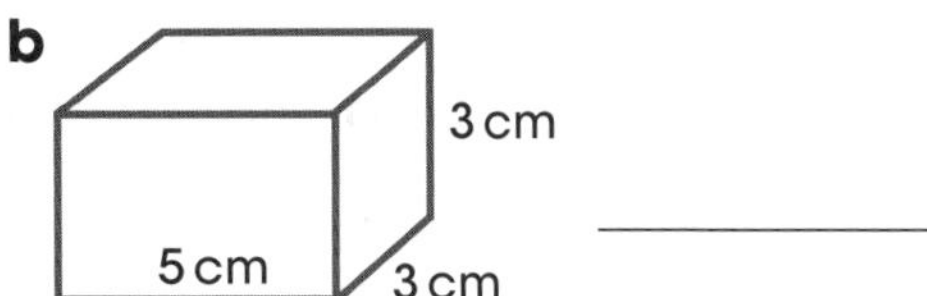

____

Fractions of Shapes

DATE:

You will need: coloured pencils

1 Shade each shape to show the fraction.

a $\frac{5}{6}$ 

b $\frac{2}{3}$ 

c $\frac{7}{8}$ 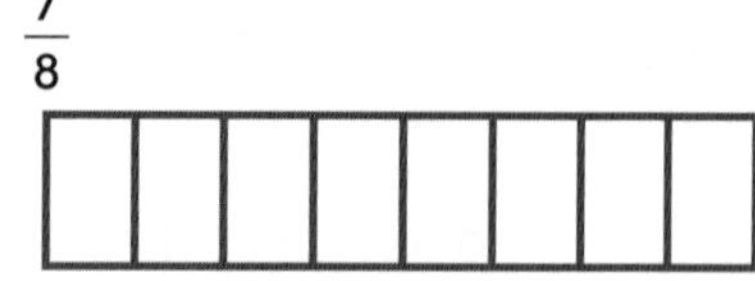

d $\frac{1}{4}$ 

e $\frac{2}{3}$ 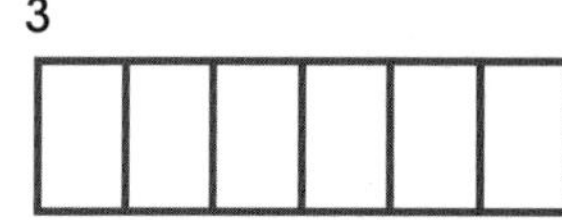

f $\frac{3}{4}$ 

2 Shade each set of circles to show the fraction.

a seven tenths   

b two thirds   

c seven eighths   

3 Write the fraction for the shaded circles in each row.

a       

b        

c  

4 Using the space below, draw each shape and shade the fraction.

a three fifths of a circle **b** four eighths of a square **c** five tenths of a rectangle

Extension: On another sheet of paper, use shapes to show the following mixed fractions.

a one and three quarters **b** two and a half **c** one and three eighths

d two and two thirds **e** three and a quarter **f** two and five sevenths

Ordering Fractions

DATE:

1 Order each set of fractions from **smallest** to **largest**.

a $\frac{5}{8}, \frac{3}{4}, \frac{1}{2}, \frac{1}{8}$ ____________________

b $\frac{1}{3}, \frac{2}{9}, \frac{2}{3}, \frac{7}{9}$ ____________________

c $\frac{5}{6}, \frac{1}{6}, \frac{2}{3}, \frac{1}{2}$ ____________________

d $\frac{1}{4}, \frac{1}{12}, \frac{5}{6}, \frac{3}{4}$ ____________________

2 Place the following fractions on the number line.

$\frac{1}{2}$ $\frac{1}{8}$ $\frac{3}{4}$ $\frac{7}{8}$ $\frac{1}{5}$ $\frac{3}{10}$ $\frac{1}{3}$ $\frac{9}{10}$

3 Draw a number line for each problem. Show the fractions on the number lines.

a quarters and sixths

b fifths and eighths

c thirds and tenths

4 On a 0° to 100° thermometer, what temperature would it be if the mercury was:

a half full? ____________ **b** three quarters full? ____________

c eight tenths full? ____________ **d** two fifths full? ____________

Extension: Draw a number line showing twelfths and twentieths.

Twin and Triplet Trouble

DATE:

There are 18 students waiting to be picked up from school, and everybody is either a twin or a triplet. People who are related to each other have **equivalent fractions** next to their names. Your job is to work out who is related to whom.

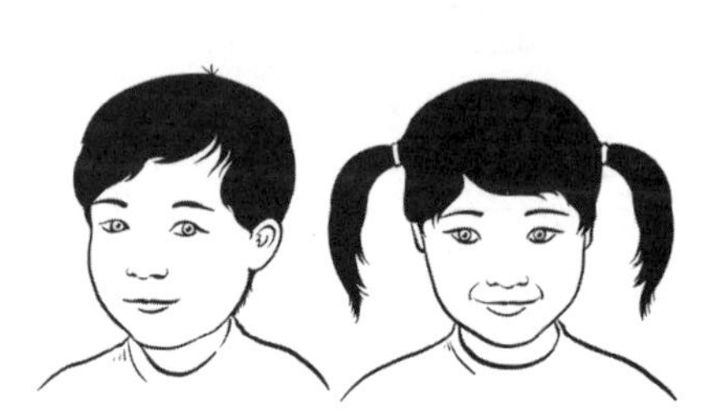

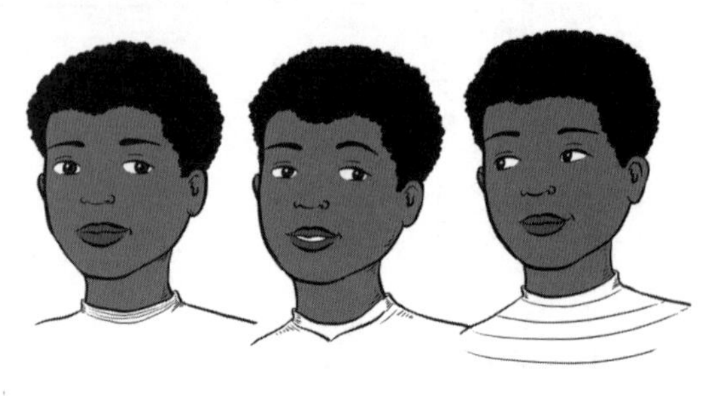

Clare	$\frac{4}{8}$	Chloe	$\frac{3}{4}$	Lia	$\frac{1}{3}$
Leo	$\frac{4}{16}$	Thomas	$\frac{6}{9}$	Emily	$\frac{8}{10}$
Makayla	$\frac{2}{10}$	Emma	$\frac{6}{12}$	Nathan	$\frac{1}{5}$
Isabella	$\frac{4}{12}$	Lily	$\frac{3}{5}$	Noah	$\frac{1}{4}$
Alicia	$\frac{4}{5}$	Miles	$\frac{3}{12}$	Lachlan	$\frac{2}{3}$
Laura	$\frac{6}{8}$	Kate	$\frac{6}{10}$	Amy	$\frac{2}{6}$

List the sets of twins and triplets below.

Twins ______________________ Twins ______________________

Twins ______________________ Twins ______________________

Twins ______________________ Twins ______________________

Triplets ______________________ Triplets ______________________

Extension: Using your knowledge of percentages, work out which kids each parent has come to pick up.

Jason: 80% ______________________ Richard: 60% ______________________

Darlene: 20% ______________________ Nicole: 66.7% ______________________

Graham: 75% ______________________ Jane: 33.3% ______________________

Donna: 50% ______________________ Patrick: 25% ______________________

DATE:

STUDENT ASSESSMENT

1 Shade each diagram to show the fraction.

a $\frac{2}{3}$ 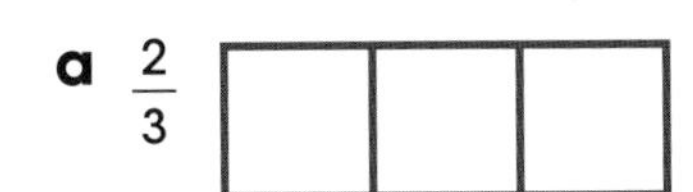

b $\frac{3}{4}$ 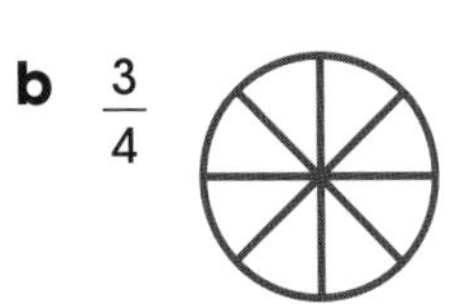

c $\frac{1}{5}$ 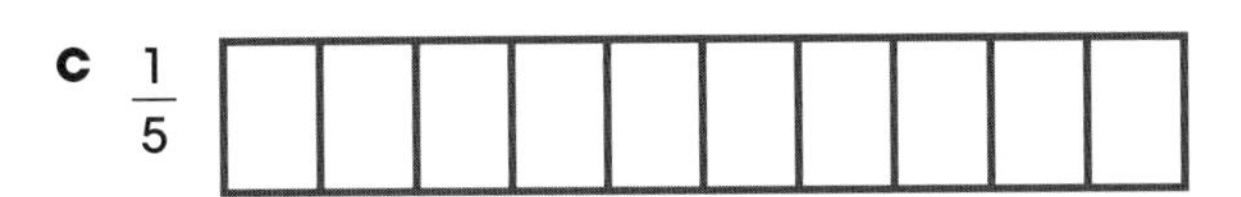

d $\frac{3}{10}$ 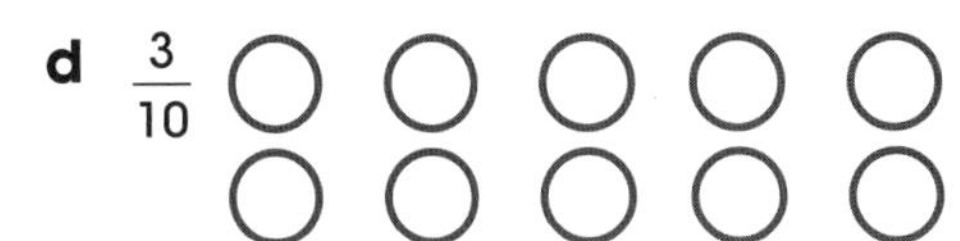

e $\frac{1}{2}$

f $\frac{1}{3}$

2 Order each set of fractions from **smallest** to **largest**.

a $\frac{1}{2}, \frac{3}{4}, \frac{2}{3}, \frac{1}{3}, 1$ ______________________________

b $\frac{3}{10}, \frac{1}{2}, \frac{4}{5}, \frac{1}{5}, \frac{3}{5}$ ______________________________

3 Find the equivalent fractions.

a $\frac{2}{3} = \frac{}{6}$

b $\frac{3}{5} = \frac{}{10}$

c $\frac{1}{2} = \frac{}{8}$

d $\frac{3}{4} = \frac{}{12}$

4 Simplify each improper fraction.

a $\frac{12}{8} =$

b $\frac{7}{5} =$

c $\frac{10}{4} =$

d $\frac{11}{3} =$

5 Alex has 24 soft toys. One third of them are cats. How many soft toy cats does Alex have?

Positive and Negative Numbers

DATE:

1 Put each set of numbers in ascending order.

a 6 10 -4 -2 -7

____

b -1 -9 5 0 -3

____

c 12 -3 9 -11 -1

____

d -9 -4 -12 -20 2

____

2 Circle the largest of each pair of numbers.

a -4 2 **b** -6 1 **c** 10 -1 **d** -20 -1 **e** 1 -9

f -8 -2 **g** 8 -10 **h** 0 -7 **i** -18 -21 **j** 31 -16

3 Use either the "greater than" symbol (>) or the "less than" symbol (<) to complete the statements below.

a 12 6 **b** 8 0 **c** -1 7 **d** -2 8 **e** 11 -6

f -9 0 **g** 18 -5 **h** 0 -9 **i** 20 -1 **j** -28 -19

4 Complete each statement.

a three less than negative nine is ______________________

b five more than negative twelve is ______________________

c four less than zero is ______________________

d nine more than negative ten is ______________________

e eight less than two is ______________________

f ten less than six is ______________________

Extension: Investigate when you would use negative numbers in real life.

Write some ideas here.

____

____

Number Lines

DATE:

1 Fill the gaps on each number line.

a

b

c
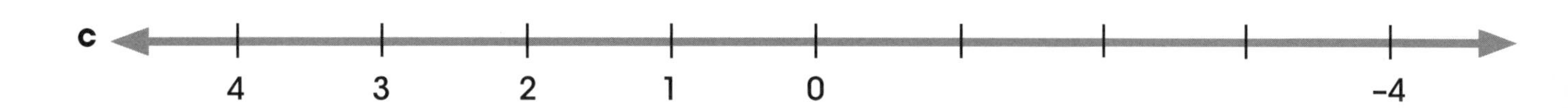

d
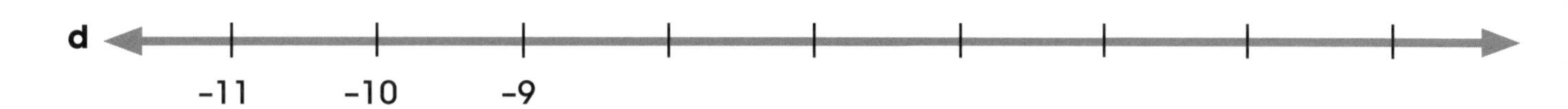

e
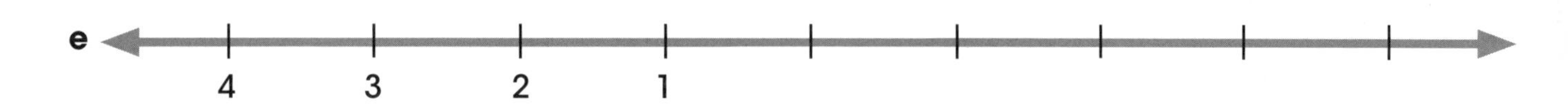

f
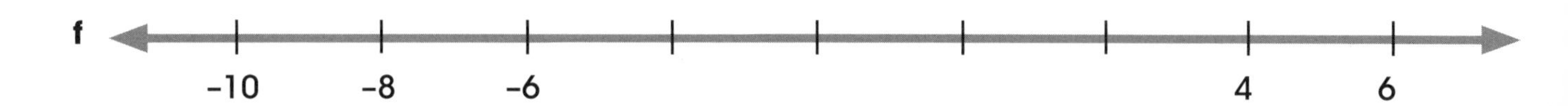

2 Complete each number pattern.

a -4, -3, -2, -1, _____, _____, _____, _____

b -6, -4, -2, _____, _____, _____, _____, _____

c 8, 5, 2, _____, _____, _____, _____, _____

d -10, -6, -2, _____, _____, _____, _____

e 5, 1, -3, _____, _____, _____, _____, _____

f 20, 14, 8, _____, _____, _____, _____, _____

3 Use the number lines to show each number pattern.

a From 0, count backwards by 4.

b From -10, count forwards by 6.

c From 6, count backwards by 3.

d From 12, count backwards by 8.

Problems with Positive and Negative Numbers

DATE:

1 Use words to complete each statement.

a six less than negative nine is ______________

b ten more than negative eleven is ______________

c twelve less than zero is ______________

d six more than negative nine is ______________

e nine more than negative five is ______________

f twenty less than five is ______________

2 Find the differences between the **minimum** and **maximum** temperatures.

a Min: 2°C
Max: 9°C ________

b Min: -1°C
Max: 9°C ________

c Min: -6°C
Max: 7°C ________

d Min: -7°C
Max: 7°C ________

e Min: -10°C
Max: 7°C ________

f Min: -4°C
Max: 12°C ________

3 **a** 8 - 12 = ________ **b** 4 - 7 = ________ **c** 1 - 9 = ________

d 11 - 15 = ________ **e** 9 - 18 = ________ **f** 13 - 16 = ________

4 In indoor cricket, players lose 5 runs every time they go out, rather than leaving the field. Each pair bats for 4 overs. Work out how many runs each pair in team Bees scored, when all the lost runs from their outs are subtracted. Add them together for the team's total score.

Team Bees	Runs	Outs	Total
Melissa and Tahli	32	3 (-15 runs)	
Phoebe and Hayley	18	5 (-25 runs)	
Jess and Abbie	14	4 (-20 runs)	
Sakura and Mei	26	4 (-20 runs)	
TEAM TOTAL			

If the other team scored 7 runs, did the Bees win or lose? By how much?

____

Extension: Roll a 10-sided dice twice, then subtract the **larger** number from the **smaller** number to get a negative answer. Keep rolling the dice and subtracting each number you roll from your total, until you get to -30. Keep a record of the numbers and calculations on a sheet of paper.

DATE:

STUDENT ASSESSMENT

1 Order the numbers from **smallest** to **largest**.

a -1, -4, 0, 8, 3, -2 ______________________________

b -3, 1, -1, 4, 3, -2 ______________________________

2 **a** Draw a number line and show these numbers: 1, -4, -2, 3, 5, -5, 2, 0

b Write 3 equations that could be solved using your number line.

____

____

____

3 Use < or > to complete the following statements.

a -1 2

b -3 3

c -4 -1

d 1 -5

4 Solve:

a -3 + 1 =

b -6 - 4 =

c -2 + 4 =

d 3 - 9 =

5 Find:

a six more than negative eight ____________________

b two less than one ____________________

c three more than negative seven ____________________

d four less than negative two ____________________

Chance with Dice

DATE:

1 If you roll a 6-sided dice, what are the chances of rolling:

a an even number?

b a number greater than 2?

c a prime number?

d a number less than 3?

e a four?

f a nine?

2 What are the chances of rolling an odd number on:

a a 6-sided dice?

b a 12-sided dice?

c a 10-sided dice?

d a 20-sided dice?

e a 100-sided dice?

f a 7-sided dice?

3 Make your own dice! For each one, look at the scale to see what the number in brackets stands for (impossible, unlikely, etc.). Choose numbers to put on the dice to match the likelihood of getting each result.

Scale	
1	impossible
2	very unlikely
3	unlikely
4	equal chance
5	likely
6	very likely
7	certain

a an even number (4)

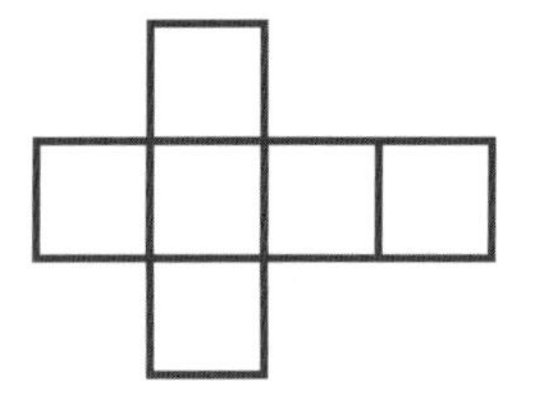

b an odd number (2)

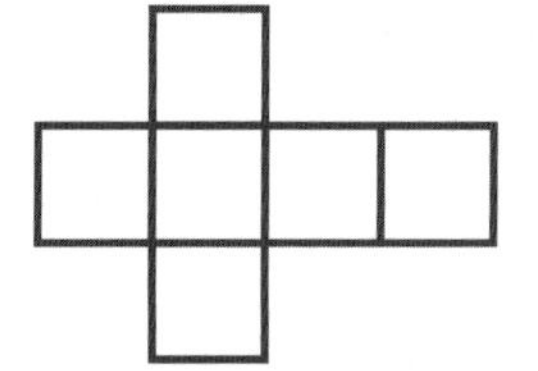

c a 2-digit number (5)

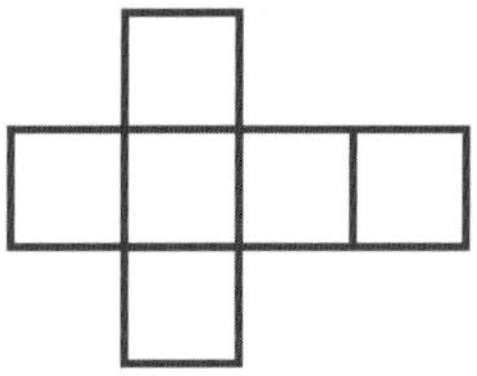

d a number less than 4 (3)

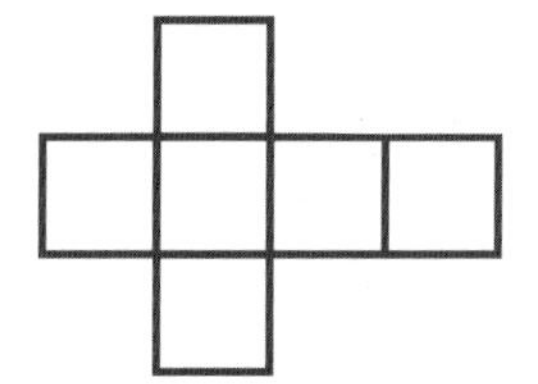

e a 6 (6)

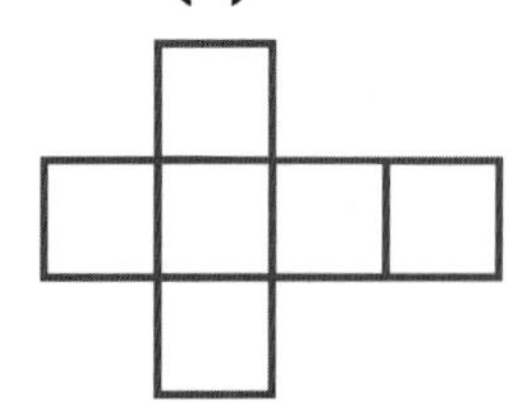

f a 3 (1)

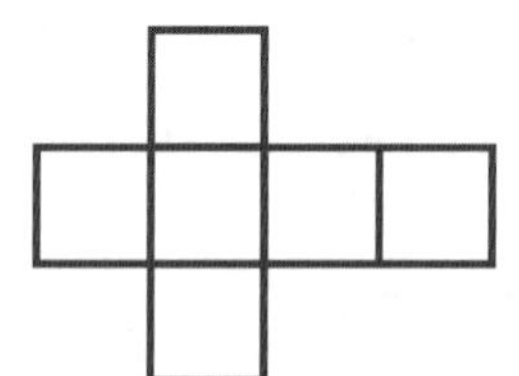

g a number less than 6 (7)

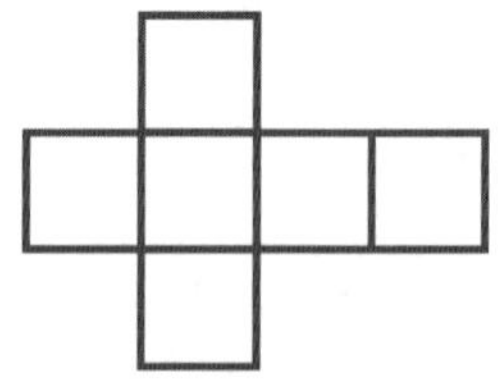

4 Using the terms in the scale, write the likelihood that the following will happen.

a The sun will rise tomorrow morning: ______________________

b The next baby born in Australia will be female: ______________________

c A unicorn is eating your lunch right now: ______________________

Extension: Use another sheet of paper. For each of the items on the scale, write down an event that matches the chance of it happening.

Chance (TRB pp. 92–95)
Chance MA3-19SP conducts chance experiments and assigns probabilities as values between 0 and 1 to describe their outcomes

Chance with Cards

DATE:

You will need: a calculator, a deck of cards

Answer the following questions using fractions.

1 If you draw a card from a full deck of cards, what is the likelihood of drawing:

a the 3 of hearts?

b a spade?

c a red card?

d a four?

e a picture card (jack, queen or king)?

f an even number?

2 Take out the picture cards. How many cards are left? ___________

3 Complete the following table, using the deck of cards without picture cards. Use a calculator to work out the decimal and percentage for each event. One has been done for you.

What are the chances of drawing:

	Fraction	Decimal	Percentage	Worded
A black card	$\frac{20}{40}$	0.5	50%	equal chance
An ace				
A card greater than 3 (aces count as 1s)				
The four of diamonds				
A king				
Anything but an ace				
An even number				

Extension: What would happen to your chances if you had 2 attempts to draw one of the cards in Question 3? Explain you answer.

____

____

Spinners

DATE:

You will need: coloured pencils

1 Colour the spinner so that the following statements are all true.

a white has a 1 in 8 chance

b blue has $\frac{1}{4}$ chance

c purple and yellow have equal chances

d orange has 1 less chance than blue

e red has no chance

f green has double the chance of white

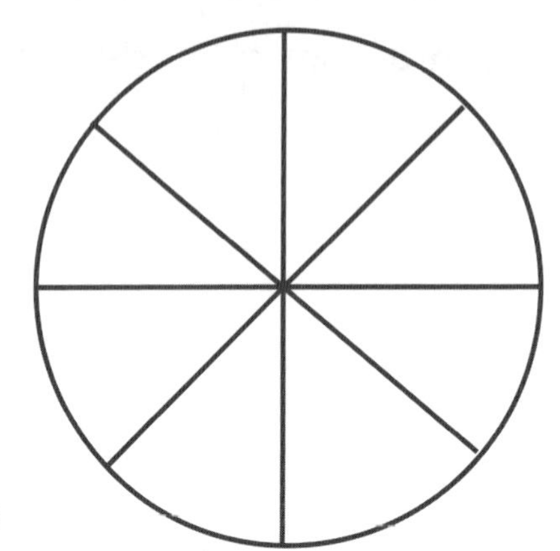

2 Look at the scale to see what each number in brackets stands for.
Shade the spinners to show that probability of the spinners landing on blue.

a 3 

b 2 

c 5 

d 7

e 4 

f 1 

	Scale
1	impossible
2	very unlikely
3	unlikely
4	equal chance
5	likely
6	very likely
7	certain

3 Which spinner has the greatest chance of landing on each of the colours below? Write the spinner letter and the probability. One has been done for you.

A B C D E F

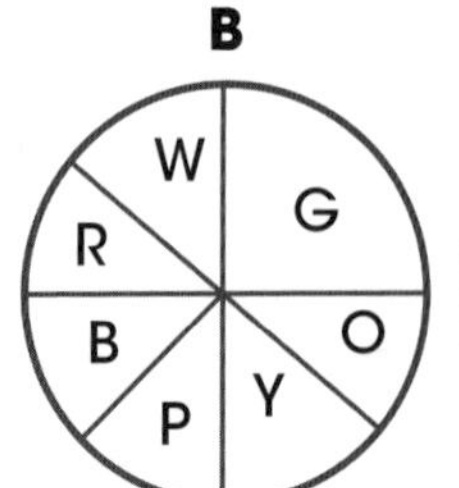

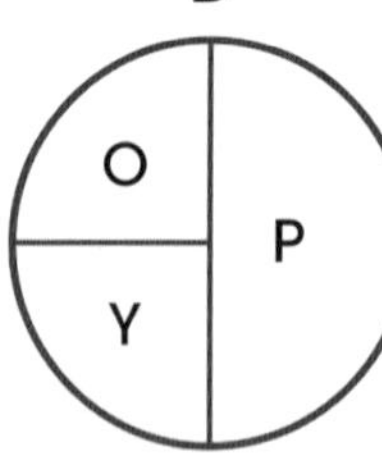

a red C $\frac{5}{8}$ ________ **b** blue ________ **c** orange ________

d green ________ **e** yellow ________ **f** purple ________

Extension: On another sheet of paper, list where you've seen something similar to a spinner in real life. What was the chance of "winning"? What was the prize?

DATE:

STUDENT ASSESSMENT

You will need: coloured pencils

1 What is the chance of:

a rolling an odd number on a 6-sided dice?

b rolling an even number on a 10-sided dice?

c drawing a red 4 from a full deck of cards?

d drawing an odd number from a full deck of cards (aces count as 1)?

2 Complete the table with the chance of rolling each result with a 6-sided dice.

	Fraction	Decimal	Percentage	Worded explanation
A 6				
An even number				
A number less than 4				

3 Colour the spinner to show:

a Red has a $\frac{1}{2}$ chance.

b Green has a $\frac{1}{4}$ chance.

c Purple and yellow have an equal chance.

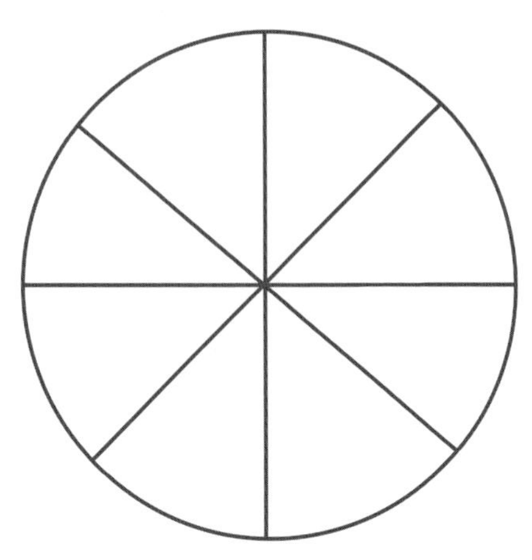

4 List 3 games that use dice and are based on chance.

a ______________________________

b ______________________________

c ______________________________

Collecting Dice Data

DATE:

You will need: a 6-sided dice

1 Roll the dice 100 times. Each time you roll, cross off the roll number in the grid on the right, to help you keep track. Use the table on the left to tally how many times each number on the dice is rolled.

	Tally	Total
1		
2		
3		
4		
5		
6		

Rolls									
1	2	3	4	5	6	7	8	9	10
11	12	13	14	15	16	17	18	19	20
21	22	23	24	25	26	27	28	29	30
31	32	33	34	35	36	37	38	39	40
41	42	43	44	45	46	47	48	49	50
51	52	53	54	55	56	57	58	59	60
61	62	63	64	65	66	67	68	69	70
71	72	73	74	75	76	77	78	79	80
81	82	83	84	85	86	87	88	89	90
91	92	93	94	95	96	97	98	99	100

2 Write down 3 interesting facts about your data.

____

____

____

3 Write down 3 questions you could ask about your data.

____

____

____

Three Coins

DATE:

You will need: 3 coins

1 Toss the 3 coins 50 times. Each time, put a tally in the table to record what combination of heads and tails you got (e.g. HHT means 2 heads and 1 tail). Keep track of your coin tosses by crossing off a number in the grid on the right for each one.

	Tally	Total
HHH		
HHT		
HTT		
TTT		

Tosses									
1	2	3	4	5	6	7	8	9	10
11	12	13	14	15	16	17	18	19	20
21	22	23	24	25	26	27	28	29	30
31	32	33	34	35	36	37	38	39	40
41	42	43	44	45	46	47	48	49	50

2 Write down any interesting facts about your data.

____

____

____

3 Is it more likely you will toss HHH or HHT, or is there an equal chance? Why?

____

____

____

4 Is it more likely to toss HHT or HTT, or is there an equal chance? Why?

____

____

____

Extension: On another sheet of paper, work out the probability of each of the 4 results, using fractions, decimals and percentages.

Collecting Card Data

DATE:

You will need: a deck of cards, a partner

1 Draw pairs of cards from a full deck, putting each pair back before you draw the next pair. Use the space below to tally how many times you draw a black and a red card, two black cards, or two red cards. You can record this data however you like.

2 Write down any interesting facts about your data.

3 Compare your data with your partner's data, using the table. How is it different?

	You	Partner
Black and black		
Black and red		
Red and red		

Extension: How many combinations of black and red are possible if you draw 5 cards at a time? What do you think the chances are of drawing all black?

Collecting Data (TRB pp. 96–99)
Chance MA3-19SP conducts chance experiments and assigns probabilities as values between 0 and 1 to describe their outcomes

STUDENT ASSESSMENT

1 Create a tally table of the data for tossing two coins 20 times in a row. Here are the results of each toss:

HH TT HT TH TH

HH HT TH TT TT

HH HT TT TT TT

HT TH HH TT HH

2 Write 3 questions about the data.

a ______________________________

b ______________________________

c ______________________________

3 Do you think the data would be the same every time you did this activity?

____

____

____

4 Give 3 examples of chance used in everyday life.

a ______________________________

b ______________________________

c ______________________________

5 List 3 activities based on chance that you could complete with a deck of cards.

a ______________________________

b ______________________________

c ______________________________

Lacrosse Number Sequences

DATE:

After a long and hard season, the lacrosse grand final has come down to the Owls and the Yaks. To find out who wins, use your knowledge of number sequences! Continue each sequence until you find the last (eighth) number.

If that number is **even**, count one goal for the Owls (O under "Goal").

If that number is **odd**, count one goal for the Yaks (Y under "Goal").

FIRST HALF

Sequence	Goal
0, 3, 6, 9, ____, ____, ____, ____	
10, 20, 30, 40, ____, ____, ____, ____	
1, 2, 4, 8, ____, ____, ____, ____	
15, 20, 25, 30, ____, ____, ____, ____	
5, 10, 20, 40, ____, ____, ____, ____	
11, 17, 23, 29, ____, ____, ____, ____	
33, 30, 27, 24, ____, ____, ____, ____	
11, 23, 35, 47, ____, ____, ____, ____	
-10, -5, 0, 5, ____, ____, ____, ____	
100, 91, 82, 73, ____, ____, ____, ____	

SECOND HALF

Sequence	Goal
3, 9, 15, 21, ____, ____, ____, ____	
88, 77, 66, 55, ____, ____, ____, ____	
7, 14, 21, 28, ____, ____, ____, ____	
384, 192, 96, 48, 24, ____, ____, ____	
1, 4, 7, 10, ____, ____, ____, ____	
128, 64, 32, 16, ____, ____, ____, ____	
1, 8, 15, 22, ____, ____, ____, ____	
9, 18, 27, 36, ____, ____, ____, ____	
-11, -8, -5, -2, 1, ____, ____, ____	
87, 78, 69, 60, ____, ____, ____, ____	

Final Score: OWLS ________ v YAKS ________

Extension: Start your own number sequence, featuring both odd and even numbers.

____ ____ ____ ____ ____ ____ ____ ____ ____ ____ ____ ____

Is the **twelfth** number odd or even? Is there any way to predict this?

____

____

Number Sequences with Decimals

DATE:

1 Complete each number pattern.

a Rule: + 8 28, 36, 44, 52, ________, ________, ________, ________

b Rule: + 14 50, 64, 78, 92, ________, ________, ________, ________

c Rule: − 8 92, 84, 76, 68, ________, ________, ________, ________

d Rule: − 24 200, 176, 152, 128, ________, ________, ________, ________

e Rule: + 31 31, 62, 93, 124, ________, ________, ________, ________

f Rule: − 9 25, 16, 7, −2, ________, ________, ________, ________

2 Complete each number pattern.

a 0.7, 7, 70, ________, ________, ________

b 10 000, 1 000, 100, ________, ________, ________

c 1.4, 14, 140, ________, ________, ________

d 7 500, 750, 75, ________, ________, ________

e 5 000 000, 500 000, 50 000, ________, ________, ________

3 Identify each rule and complete the number patterns.

a 2.8, 4.0, 5.2, 6.4, ________, ________, ________, ________ Rule: ________

b 0.9, 1.6, 2.3, 3.0, ________, ________, ________, ________ Rule: ________

c 15.1, 14.0, 12.9, 11.8, ________, ________, ________, ________ Rule: ________

d 5.10, 4.75, 4.40, 4.05, ________, ________, ________, ________ Rule: ________

e −22, −14, −6, 2, ________, ________, ________, ________ Rule: ________

f 8, 2, −4, -10, ________, ________, ________, ________ Rule: ________

4 Make your own number rule from 0 to 6.4.

Rule: ________ 0, ________________________________ 6.4

Extension: Make your own number rule from 0 to 10. It must include thousandths.

Rule: ________

0, ____ 10

Number Sequences with Fractions

DATE:

1 Complete each number sequence.

a Rule: $+\frac{1}{2}$ $\frac{1}{2}$, 1, $1\frac{1}{2}$, 2, $2\frac{1}{2}$, ______, ______, ______, ______

b Rule: $+\frac{1}{3}$ $1\frac{1}{3}$, $1\frac{2}{3}$, 2, $2\frac{1}{3}$, ______, ______, ______, ______

c Rule: $-\frac{1}{8}$ 1, $\frac{7}{8}$, $\frac{3}{4}$, $\frac{5}{8}$, ______, ______, ______, ______

d Rule: $+\frac{1}{4}$ $\frac{1}{4}$, $\frac{1}{2}$, $\frac{3}{4}$, 1, ______, ______, ______, ______

2 Identify each rule and complete the number sequences.

a $4\frac{1}{2}$, 4, $3\frac{1}{2}$, 3, ______, ______, ______, ______ Rule: ______

b $2\frac{2}{3}$, 3, $3\frac{1}{3}$, $3\frac{2}{3}$, ______, ______, ______, ______ Rule: ______

c $\frac{1}{8}$, $\frac{1}{4}$, $\frac{3}{8}$, $\frac{1}{2}$, $\frac{5}{8}$, ______, ______, ______ Rule: ______

d 7, $6\frac{1}{3}$, $5\frac{2}{3}$, 5, ______, ______, ______, ______ Rule: ______

e $\frac{1}{10}$, $\frac{2}{10}$, $\frac{3}{10}$, ______, ______, ______, ______ Rule: ______

3 Write the number sequence for each rule. Simplify the fractions if you can.

a Rule: $+\frac{1}{8}$ 0, ______, ______, ______, ______

b Rule: $-\frac{1}{4}$ 2, ______, ______, ______, ______

c Rule: $+\frac{1}{6}$ 5, ______, ______, ______, ______

d Rule: $-\frac{2}{5}$ 8, ______, ______, ______, ______

e Rule: $+\frac{3}{4}$ $\frac{3}{4}$, ______, ______, ______, ______

4 Make your own number rule from 0 to 5. It must include fractions.

Rule: ______ 0, ______________________________ 5

Extension: On another sheet of paper, make number sequences for the following rules: $+\frac{2}{7}$, $+\frac{3}{10}$ and $-\frac{2}{5}$. Choose a number to start each sequence. Simplify the fractions when you can.

DATE:

STUDENT ASSESSMENT

1 Complete each number sequence.

a 44, 40, 36, 32, ________, ________, ________, ________

b 12, 24, 36, 48, ________, ________, ________, ________

c –10, –8, –6, –4, ________, ________, ________, ________

d 800, 798, 796, 794, ________, ________, ________, ________

2 Complete each number sequence.

a 1.5, 15, 150, ________, ________, ________, ________

b 600 000, 60 000, 6 000, ________, ________, ________, ________

3 Write the rule for each number sequence.

a 1.5, 1.7, 1.9, 2.1 Rule: ________

b –20, –15, –10, –5 Rule: ________

c $\frac{1}{4}$, $\frac{3}{4}$, $1\frac{1}{4}$, $1\frac{3}{4}$ Rule: ________

d 500, 520, 540, 560 Rule: ________

4 Create your own number sequences and state their rules.

a whole numbers increasing

____

Rule: ________

b negative numbers

____

Rule: ________

c decimal numbers

____

Rule: ________

d fractions

____

Rule: ________

Naming Angles

DATE:

You will need: coloured pencils

1 Name each angle either **acute** or **obtuse**.

a

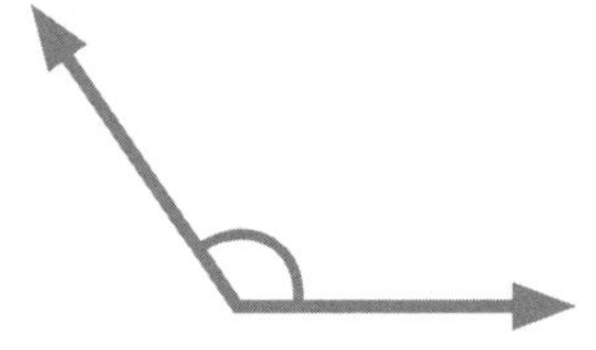

____

b

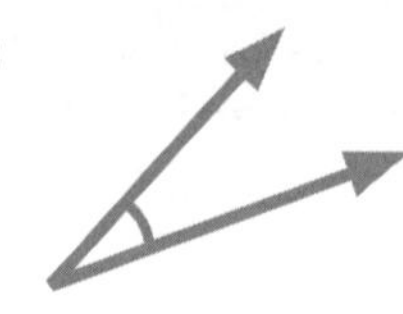

____

c

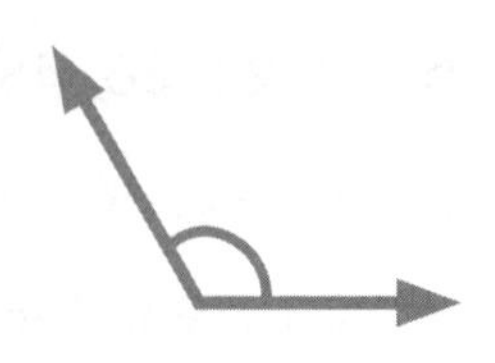

____

d

____

e

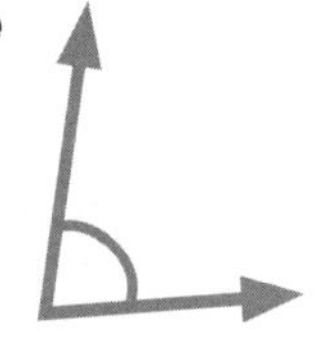

____

f

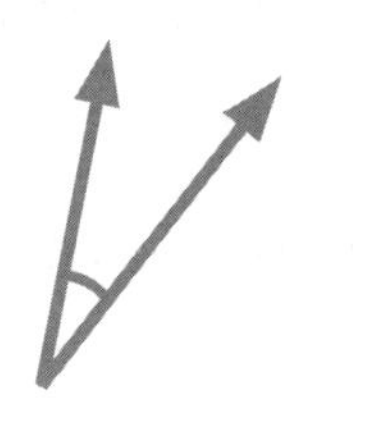

____

2 Circle the **reflex** angles.

a

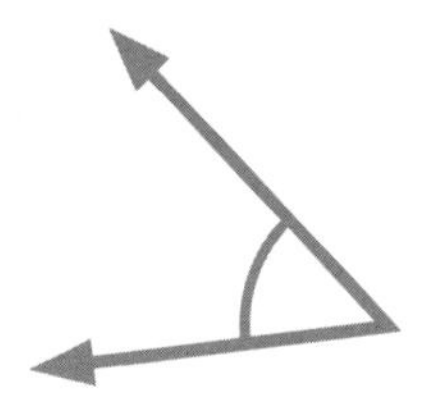

b

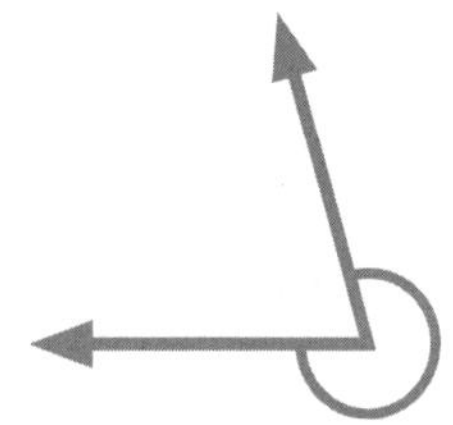

c

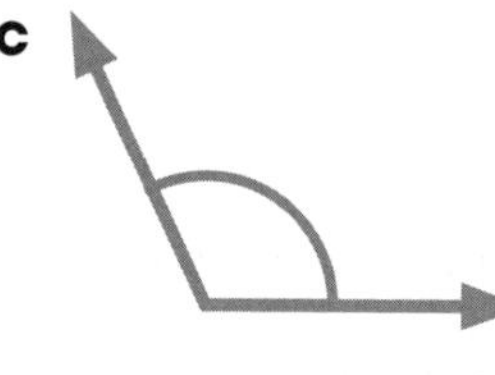

d

3 Draw and name the angles that appear in these capital letters.

W E A L T H

4 In the diagram below, colour right angles in blue, acute angles in green and obtuse angles in red.

Extension: Find examples of acute, right and obtuse angles within your room. Try to find as many of each as you can!

Measuring and Drawing Angles

DATE:

You will need: a protractor

1 Using a protractor, measure and label each angle as acute, obtuse or reflex.

a
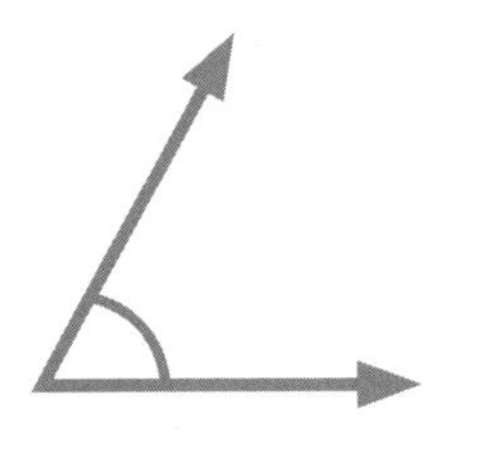

b
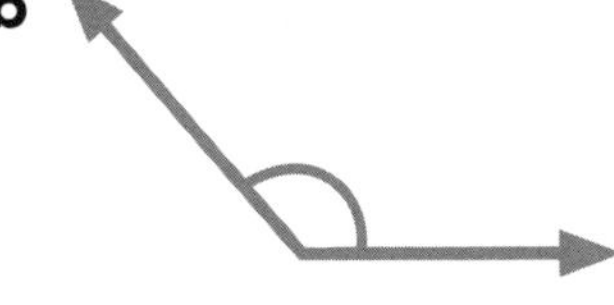

c
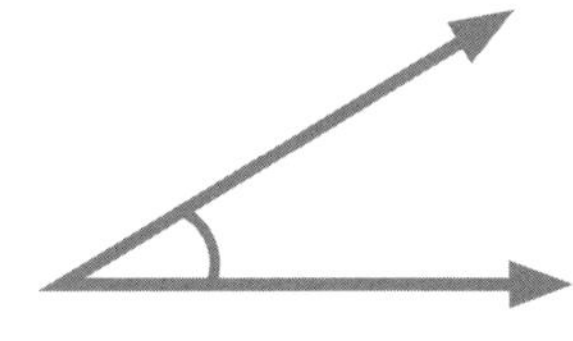

d
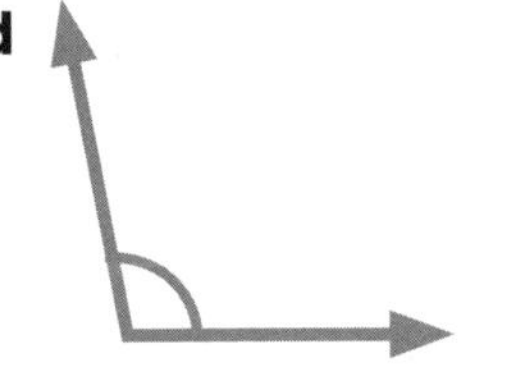

e

f
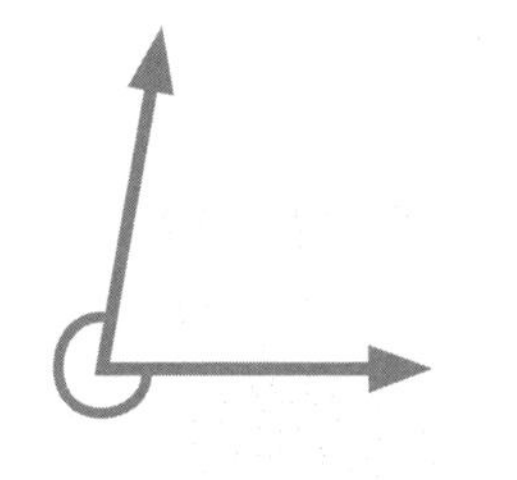

2 Use a protractor to draw the following angles.

a 45° **b** 60° **c** 20°

d 135° **e** 180° **f** 150°

3 Now, try drawing these angles **without** using a protractor.
Then use a protractor to check to see how close you were.

a 45° **b** 60° **c** 20°

Extension: On another sheet of paper, work out the **opposite** angle for each of the angles in Question 2. If the angle is acute or obtuse, the opposite angle will be reflex.

Missing Angles

DATE:

1 Find the value of each letter, without using a protractor.

a

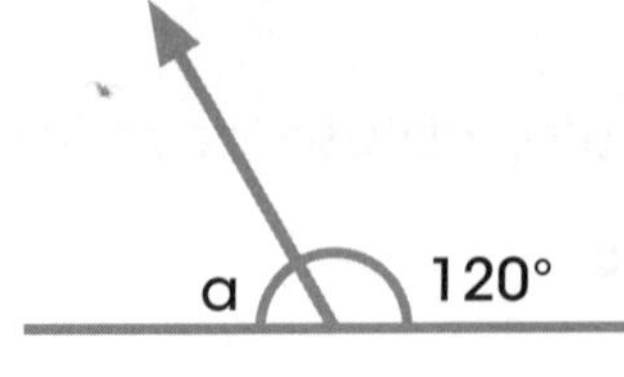

a = ________

b

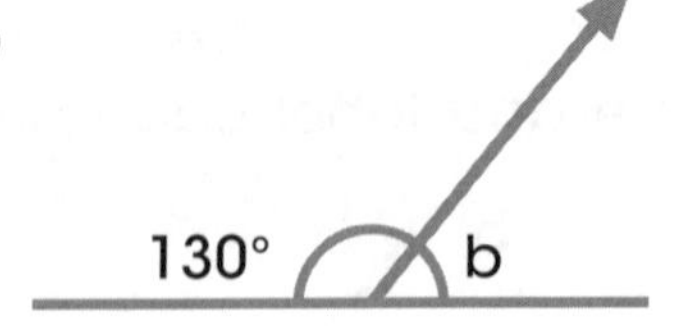

b = ________

c

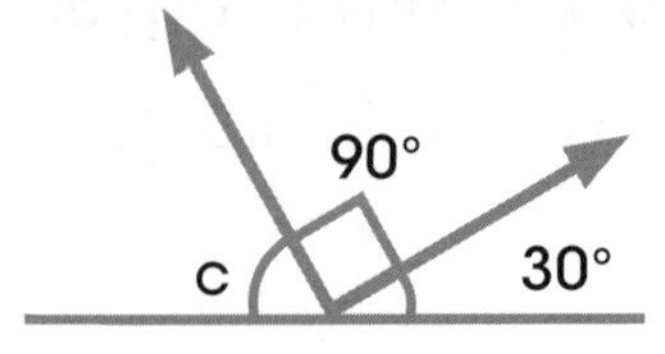

c = ________

2 Find the value of each letter, without using a protractor.

a

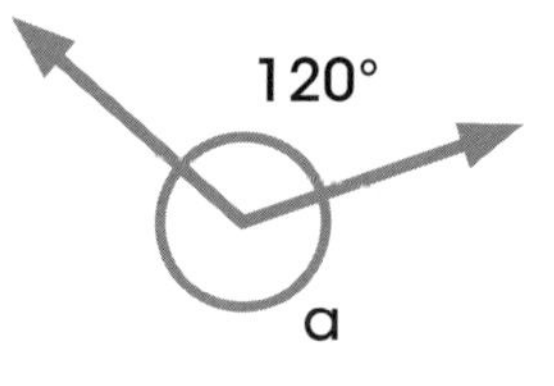

a = ________

b

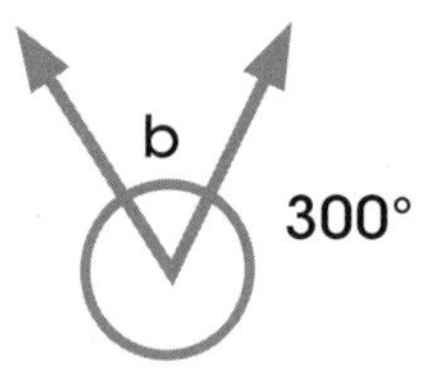

b = ________

c

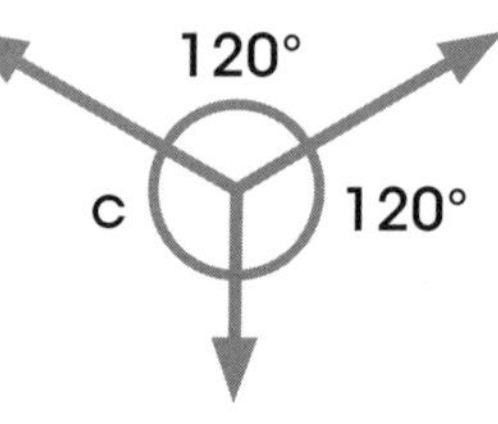

c = ________

d

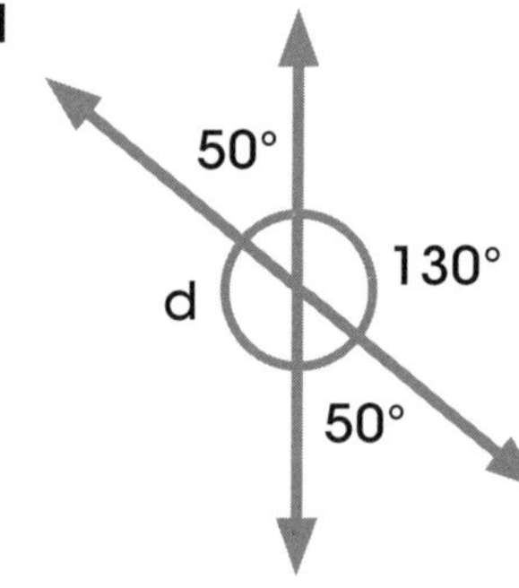

d = ________

e

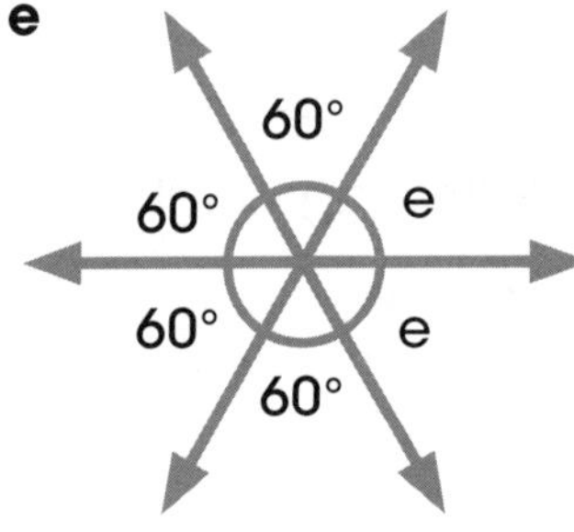

e = ________

f

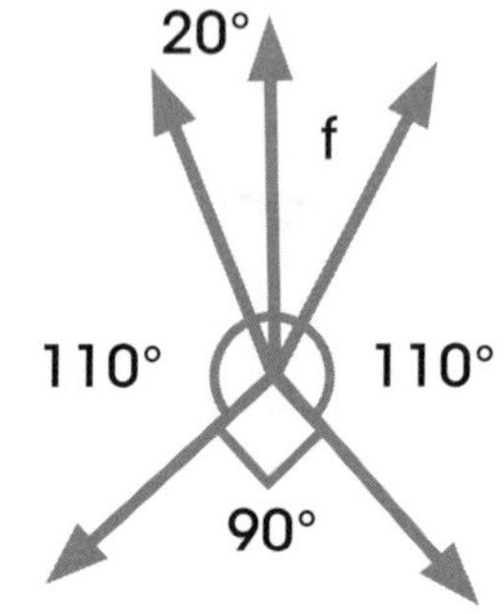

f = ________

3 Find the value of each letter, without using a protractor.

a

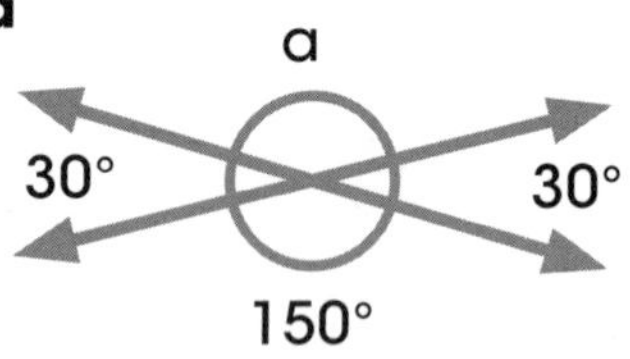

a = ________

b

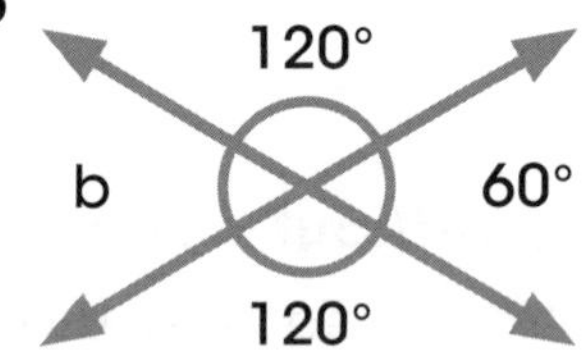

b = ________

c

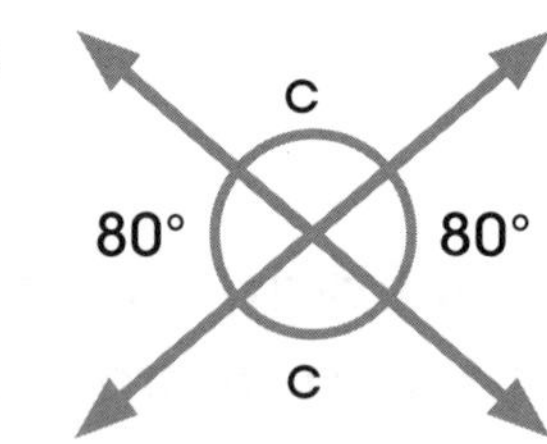

c = ________

4 In the space below, write a report on the angles you found in Question 3.

____

____

____

____

DATE:

STUDENT ASSESSMENT

1 Label each angle as acute, obtuse or reflex.

a

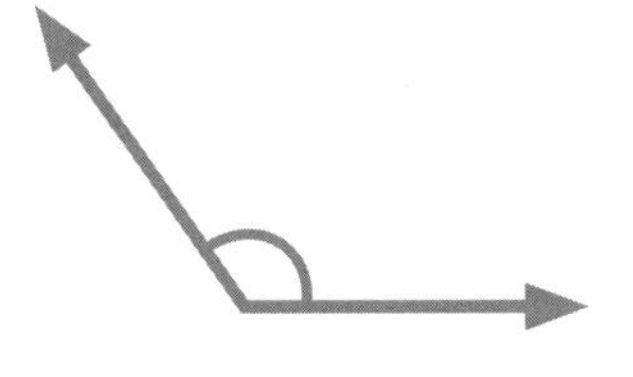

____

b

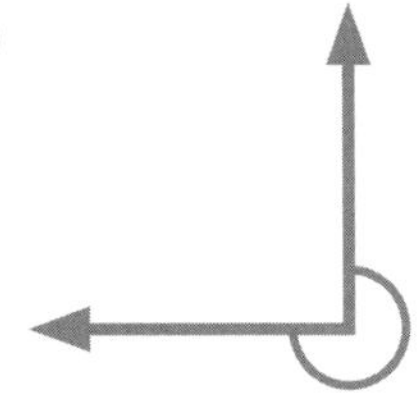

____

c

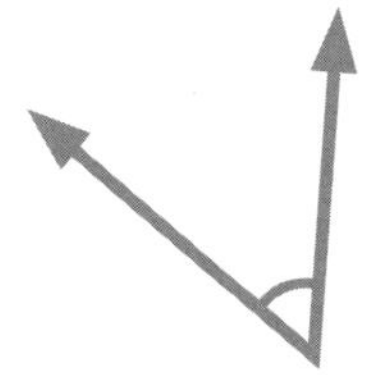

____

2 Sketch each angle.

a 60°

b 150°

c 300°

3 Find the value of each letter, without using a protractor.

a

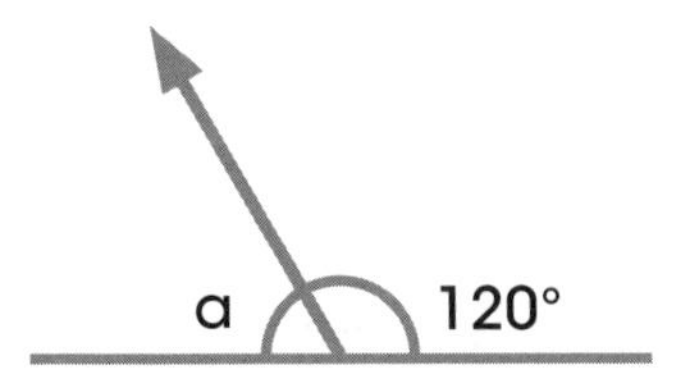

a = ________

b

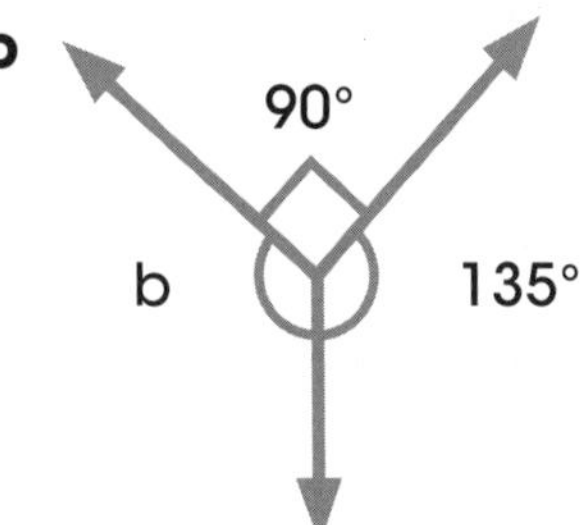

b = ________

c

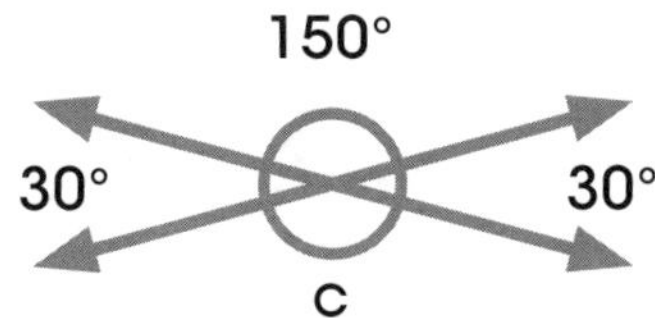

c = 

d

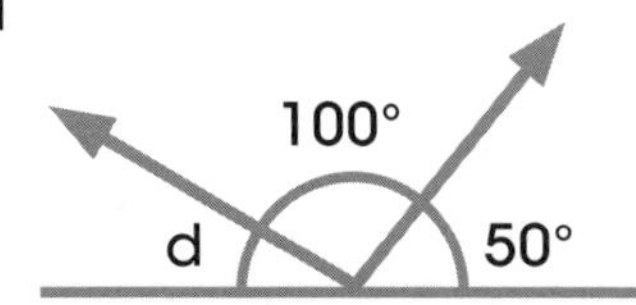

d = ________

4 Explain how you worked out the answer to Question 3d.

____

____

Addition of Fractions

DATE:

You will need: coloured pencils

1 Use the diagrams to help you add these fractions.

a $\frac{1}{5} + \frac{3}{5} =$

b $\frac{3}{8} + \frac{4}{8} =$

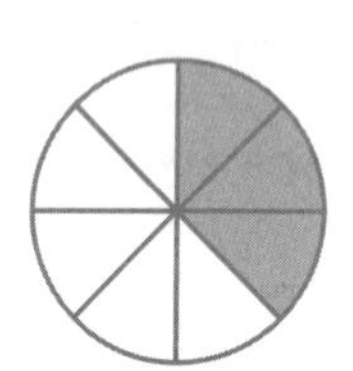

c $\frac{1}{4} + \frac{1}{2} =$

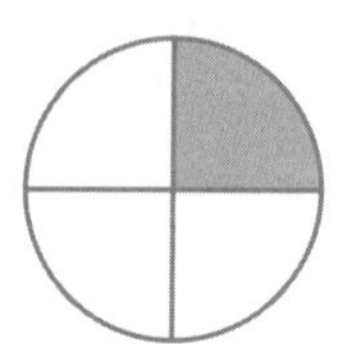

d $\frac{4}{5} + \frac{3}{5} =$

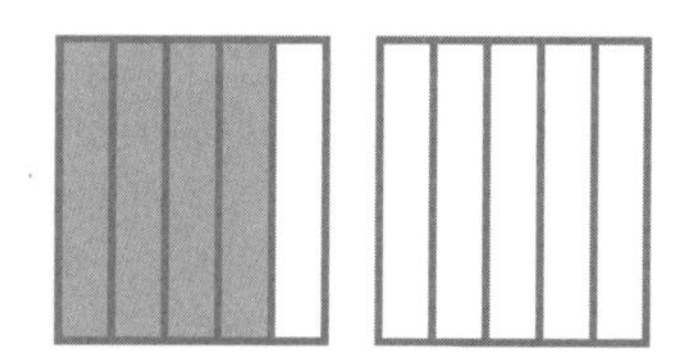

e $\frac{3}{7} + \frac{5}{7} =$

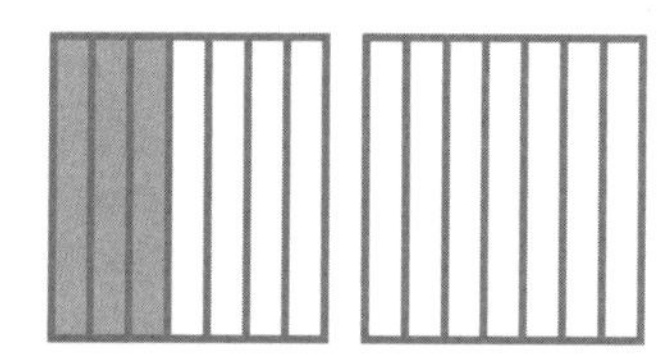

f $\frac{5}{8} + \frac{1}{2} =$

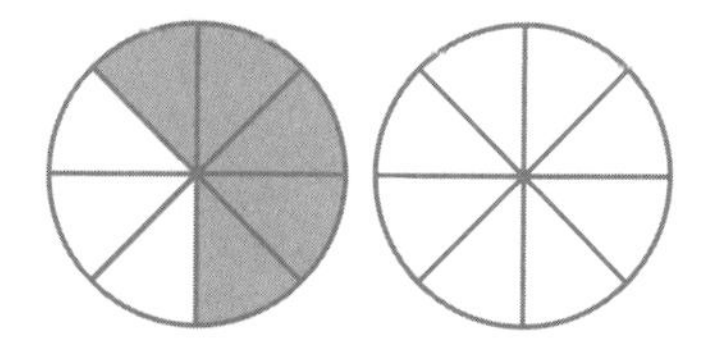

g $\frac{7}{10} + \frac{1}{5} =$

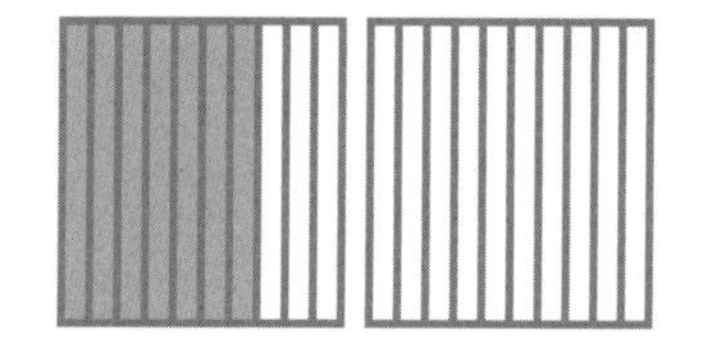

h $\frac{1}{3} + \frac{5}{9} =$

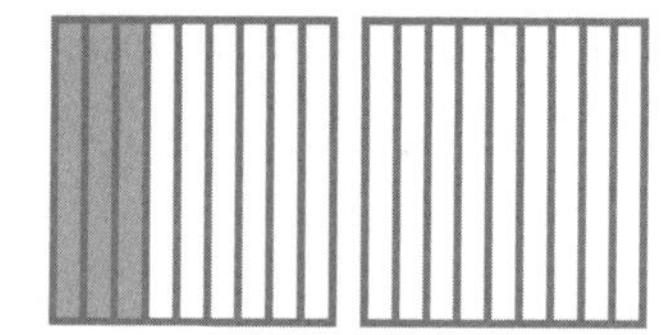

i $\frac{2}{3} + \frac{5}{6} =$

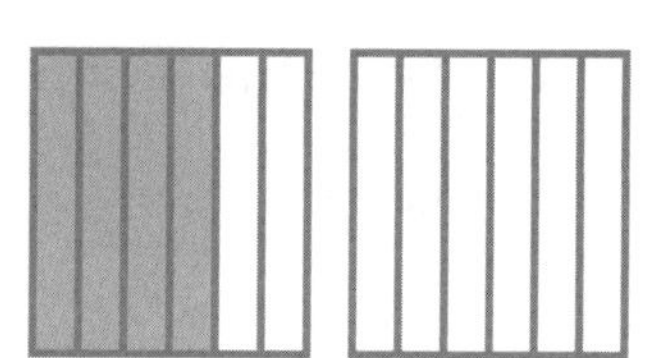

2 Show how you could use diagrams to add the fractions below, then find the answers.

a $\frac{3}{8} + \frac{3}{8} =$

b $\frac{1}{5} + \frac{7}{10} =$

c $\frac{1}{4} + \frac{5}{8} =$

d $\frac{1}{2} + \frac{3}{10} =$

e $\frac{7}{8} + \frac{1}{4} =$

f $\frac{2}{3} + \frac{1}{6} =$

Extension: On another sheet of paper, add the following mixed fractions.

a $1\frac{1}{3} + \frac{2}{3} =$

b $2\frac{2}{5} + 1\frac{3}{10} =$

c $3\frac{1}{8} + 1\frac{1}{2} =$

Fraction Addition with Number Lines

DATE:

1 Use the number lines to help you add these fractions.

a $\frac{1}{4} + \frac{3}{4} =$

b $\frac{1}{8} + \frac{5}{8} =$

c $\frac{2}{3} + \frac{2}{3} =$

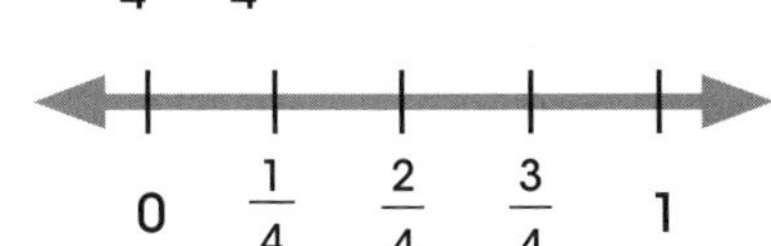

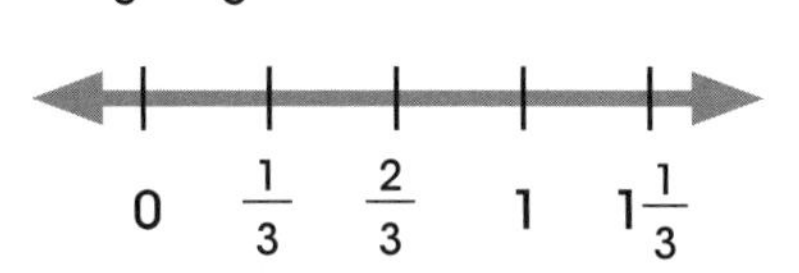

d $4\frac{1}{2} + 2\frac{1}{2} =$

e $1\frac{1}{8} + \frac{6}{8} =$

f $\frac{5}{6} + \frac{5}{6} =$

g $3\frac{1}{3} + 1\frac{2}{3} =$

h $1\frac{3}{8} + 2\frac{1}{2} =$

i $3\frac{2}{3} + 1\frac{5}{6} =$

2 Draw your own number lines to help you solve these equations.

a $2\frac{1}{2} + 1\frac{1}{2} =$

b $3\frac{5}{6} + 1\frac{3}{6} =$

c $\frac{7}{8} + \frac{6}{8} =$

d $2\frac{1}{4} + 2\frac{1}{8} =$

e $1\frac{5}{7} + 3\frac{3}{7} =$

f $2\frac{1}{2} + 2\frac{3}{10} =$

Extension: Use number lines to add the following mixed fractions.

a $4\frac{1}{2} + 3\frac{5}{8} =$

b $2\frac{1}{5} + 1\frac{7}{10} =$

c $3\frac{1}{3} + 1\frac{7}{9} =$

Fraction Race

DATE:

Jahid and Jamila are racing skateboards over a 1-kilometre track. They can only move a certain fraction of a kilometre each minute. Use the number lines below as race tracks and see who wins the race!

Jahid

Jahid skated $\frac{1}{6}$ of a kilometre in the first minute. Then he gained speed and skated $\frac{3}{12}$ in the next minute. He came across some gravel on the track and only skated $\frac{1}{12}$ of a kilometre in the minute after that. He recovered and skated $\frac{1}{6}$ of a kilometre in the next minute. He put in one last big effort and skated $\frac{1}{4}$ of a kilometre in the last minute.

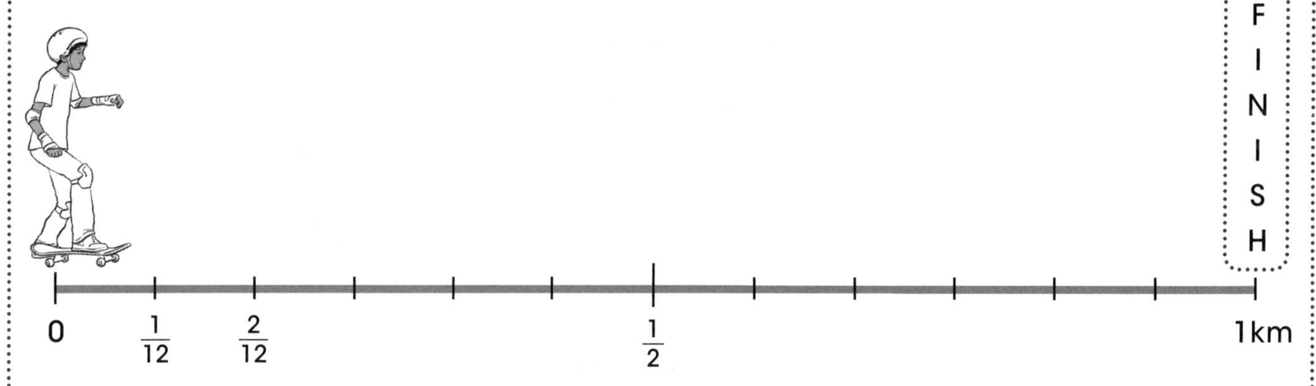

Jamila

Jamila started strong, skating $\frac{1}{4}$ of a kilometre in the first minute. She slowed going up a hill, skating $\frac{1}{12}$ of kilometre in the next minute. The minute after that, she went down the hill, travelling $\frac{3}{12}$ of a kilometre. She slowed a little to gain some control, moving $\frac{1}{12}$ of a kilometre in the next minute. Then she powered home by skating $\frac{1}{3}$ of a kilometre in the last minute.

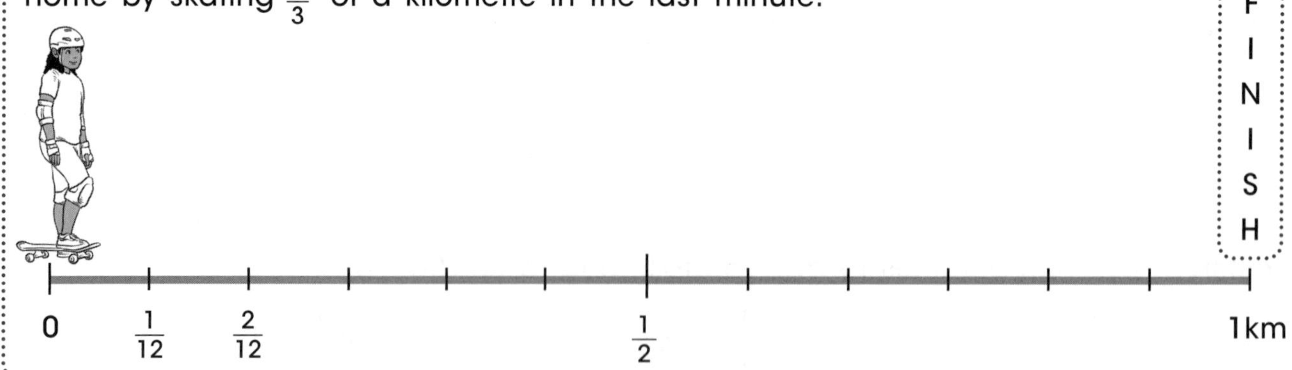

Who won the race? ________ Was the winner very far ahead? ____________________

Extension: On another sheet of paper, make your own number line from 0 to 5, and separate it into eighths. Race a partner by rolling dice. The number you roll is how many eighths you can move forward. The first one to 5 wins!

STUDENT ASSESSMENT

1 Use the diagrams to help you add these fractions.

a $\frac{2}{5} + \frac{2}{5} =$

b $\frac{5}{10} + \frac{1}{5} =$

c $\frac{7}{8} + \frac{1}{2} =$

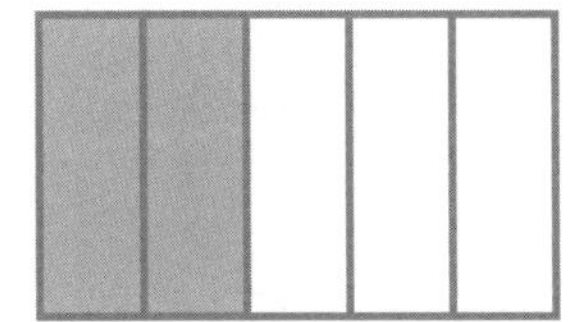

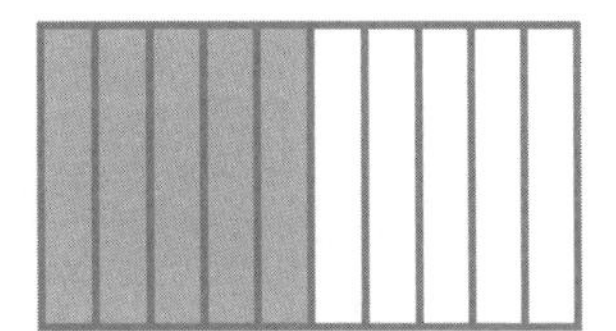

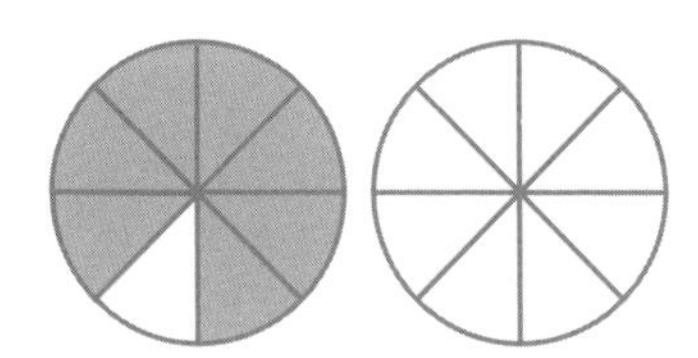

2 Add the fractions.

a $\frac{1}{3} + \frac{1}{6} =$

b $\frac{2}{3} + \frac{1}{6} =$

c $1\frac{1}{3} + \frac{4}{5} =$

d $\frac{5}{6} + \frac{1}{2} =$

3 Draw a diagram to show:

$\frac{3}{4} + \frac{1}{8} =$

4 Draw a number line to show:

$2\frac{1}{2} + 1\frac{4}{5} =$

5 Julie, the art teacher, has a bucket for water. She fills it to $\frac{1}{4}$ with water for clay, another $\frac{1}{3}$ with water for paint, and another $\frac{1}{6}$ with water for cleaning. How full is the bucket?

Subtraction of Fractions

DATE:

You will need: coloured pencils

1 Use the diagrams to help you subtract these fractions.

a $\frac{7}{8} - \frac{3}{8} =$

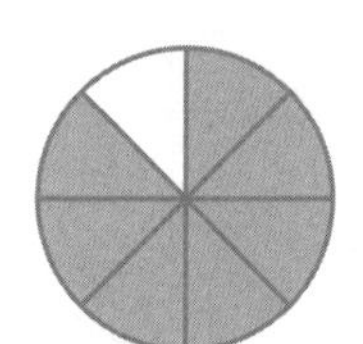

b $\frac{4}{5} - \frac{2}{5} =$

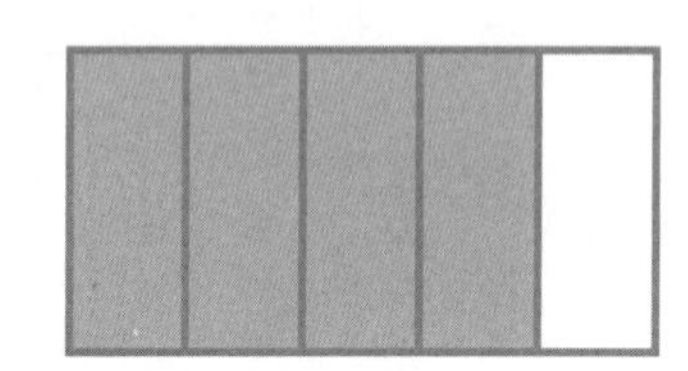

c $\frac{3}{4} - \frac{1}{8} =$

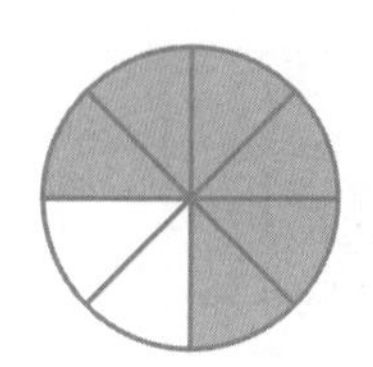

d $\frac{8}{10} - \frac{2}{5} =$

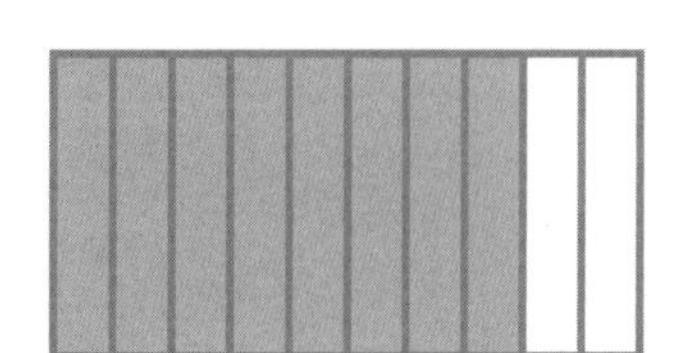

e $1\frac{1}{2} - \frac{3}{8} =$

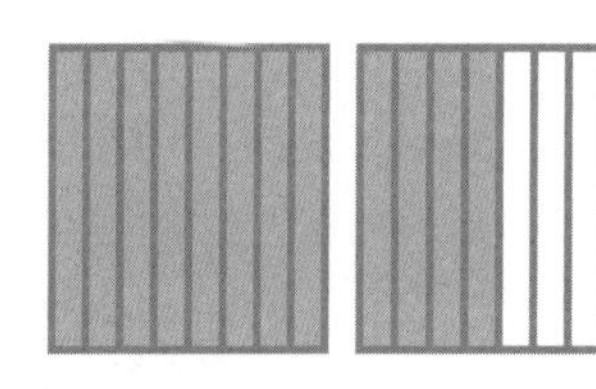

f $2\frac{1}{4} - 1\frac{1}{2} =$

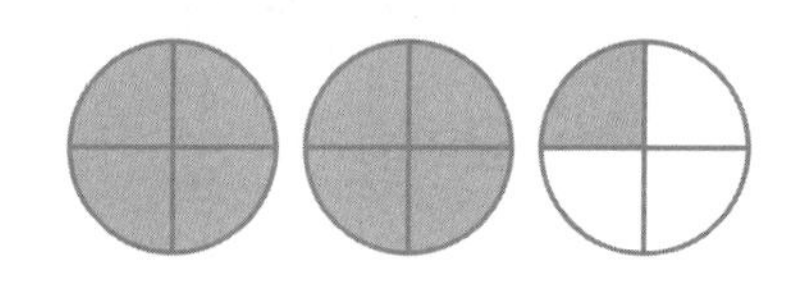

g $1\frac{6}{8} - 1\frac{3}{8} =$

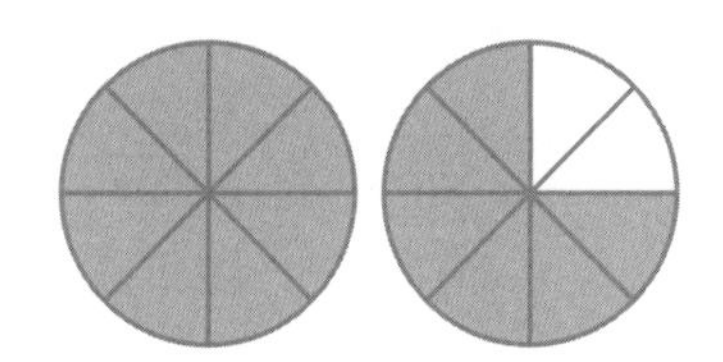

h $1\frac{7}{10} - \frac{9}{10} =$

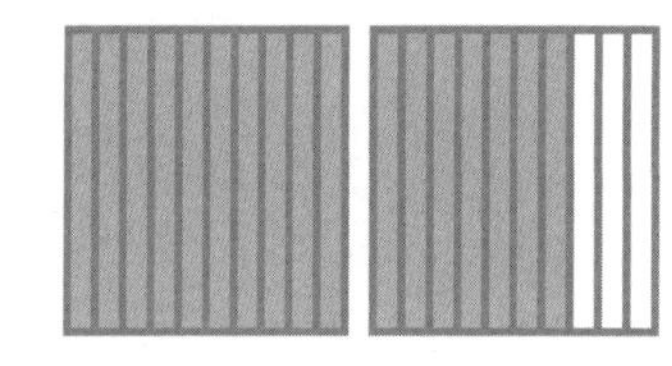

i $3 - \frac{13}{8} =$

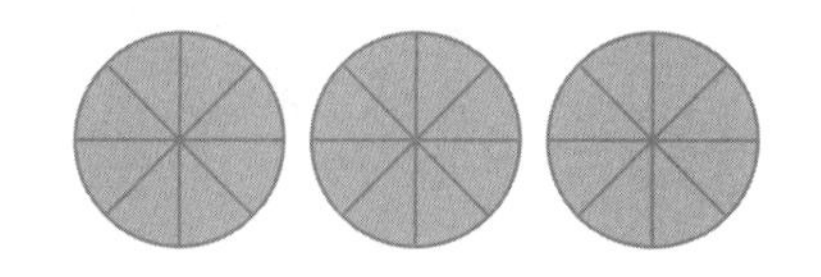

2 In the space below, use diagrams to complete the subtraction equations.

a $\frac{9}{10} - \frac{3}{10} =$

b $\frac{7}{8} - \frac{1}{4} =$

c $1\frac{2}{3} - \frac{3}{6} =$

d $2\frac{1}{2} - \frac{3}{8} =$

e $1\frac{7}{10} - 1\frac{1}{2} =$

f $\frac{11}{12} - \frac{1}{4} =$

Extension: On another sheet of paper, subtract these mixed fractions using diagrams.

a $1\frac{1}{8} - \frac{5}{8} =$

b $1\frac{2}{5} - \frac{7}{10} =$

c $2\frac{1}{8} - 1\frac{1}{2} =$

Fraction Subtraction with Number Lines

DATE:

1 Use the number lines to help you subtract the fractions.

a $1\frac{1}{2} - \frac{3}{4} =$

0 1 2

b $1\frac{5}{6} - \frac{2}{3} =$

c $\frac{7}{8} - \frac{1}{2} =$

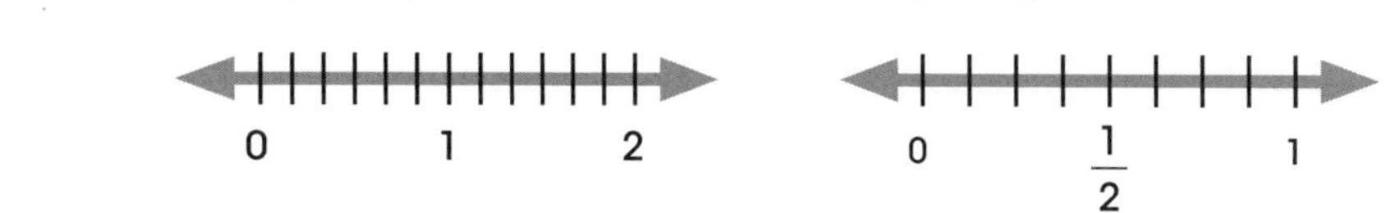

d $\frac{9}{10} - \frac{1}{5} =$

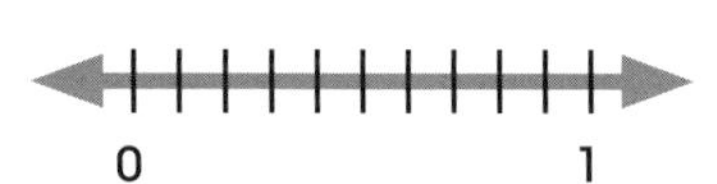

e $\frac{4}{5} - \frac{3}{10} =$

f $1\frac{3}{4} - 1\frac{1}{2} =$

g $\frac{12}{12} - \frac{3}{4} =$

h $3\frac{3}{4} - 2\frac{7}{8} =$

i $1\frac{5}{10} - 1\frac{1}{4} =$

2 Draw your own number lines to help you solve these equations.

a $2\frac{1}{2} - 1\frac{7}{8} =$

b $1\frac{2}{4} - \frac{3}{4} =$

c $2\frac{3}{8} - 1\frac{1}{2} =$

d $1\frac{7}{10} - 1\frac{1}{5} =$

e $3\frac{1}{2} - 1\frac{1}{8} =$

f $2\frac{9}{10} - 1\frac{3}{5} =$

Extension: Subtract these mixed fractions using number lines.

a $2\frac{1}{4} - 1\frac{3}{4} =$

b $2\frac{1}{5} - \frac{7}{10} =$

c $1\frac{1}{2} - \frac{7}{8} =$

Fraction Subtraction with Worded Problems

DATE:

1 Solve the equations.

a $\frac{7}{12} - \frac{3}{12} =$

b $\frac{6}{7} - \frac{1}{2} =$

c $\frac{7}{8} - \frac{1}{4} =$

d $\frac{9}{10} - \frac{2}{5} =$

e $\frac{3}{4} - \frac{1}{8} =$

f $\frac{4}{5} - \frac{3}{10} =$

g $\frac{12}{10} - \frac{2}{5} =$

h $\frac{10}{8} - \frac{1}{2} =$

i $\frac{14}{6} - \frac{5}{3} =$

2 Solve the mixed fraction equations.

a $1\frac{1}{4} - \frac{5}{8} =$

b $2\frac{3}{4} - 1\frac{1}{8} =$

c $2\frac{1}{2} - \frac{5}{6} =$

d $1\frac{1}{3} - \frac{5}{9} =$

e $2\frac{1}{10} - 1\frac{3}{5} =$

f $1\frac{7}{9} - \frac{2}{3} =$

g $3\frac{6}{7} - 1\frac{1}{7} =$

h $4\frac{1}{5} - 3\frac{6}{10} =$

i $1\frac{2}{9} - \frac{2}{3} =$

3 Liam has $1\frac{3}{4}$ kilograms of flour in the cupboard. If he uses $\frac{1}{2}$ a kilogram in a cake mix, how much will be left?

4 When Katie arrived late for dinner, there were $1\frac{3}{8}$ pizzas left. She ate half a pizza. How much pizza was left when she was finished?

5 Hamish had $\frac{6}{10}$ of a litre of milk. He drank $\frac{1}{5}$ of a litre. How much was left?

6 Rohit's guinea pig, Fuzzy, weighed $3\frac{1}{8}$ kg. Rohit put Fuzzy on a diet, making him lose $\frac{3}{4}$ of a kilogram. What does Fuzzy weigh now?

Extension: On another sheet of paper, use number lines to find the following.

a $14\frac{7}{9} - 11\frac{2}{3} =$

b $12\frac{1}{5} - 5\frac{3}{10} =$

c $23\frac{1}{2} - 11\frac{7}{12} =$

DATE:

STUDENT ASSESSMENT

1 Subtract the fractions.

a $\frac{5}{6} - \frac{1}{6} =$

b $\frac{2}{3} - \frac{1}{4} =$

c $\frac{9}{10} - \frac{3}{5} =$

d $\frac{1}{2} - \frac{3}{8} =$

2 Find:

a $1 - \frac{2}{5} =$

b $3 - 1\frac{2}{3} =$

3 Draw a number line to solve:

$3\frac{3}{8} - 1\frac{1}{4} =$

4 Draw a diagram to solve:

$1\frac{1}{2} - \frac{1}{8} =$

5 In the classroom there was a pile of counters on the bench.
If group:

A took $\frac{1}{8}$ of the counters

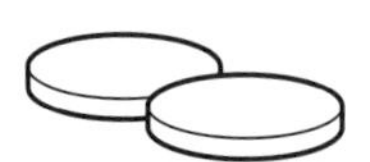

B took $\frac{1}{4}$ of the counters

C took $\frac{1}{2}$ of the counters

How many counters were left?

Analogue Time

DATE:

1 Write each time in words.

a 

b 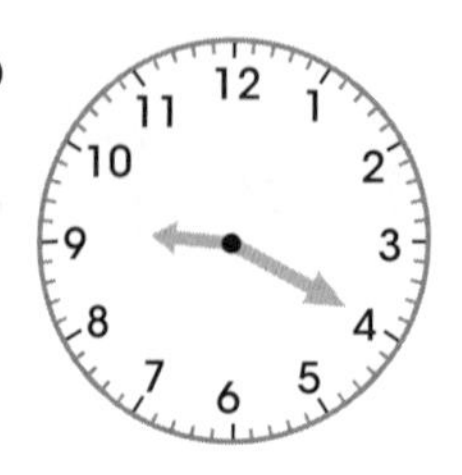

c 

d 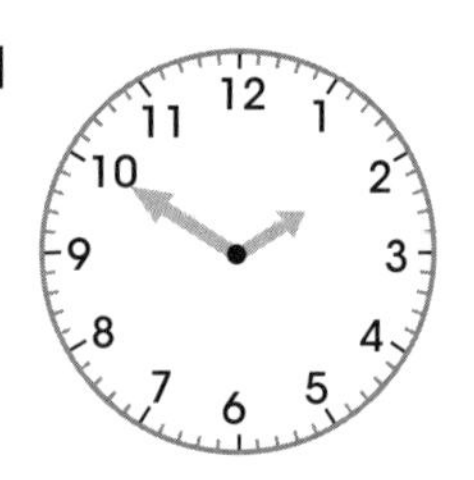

e 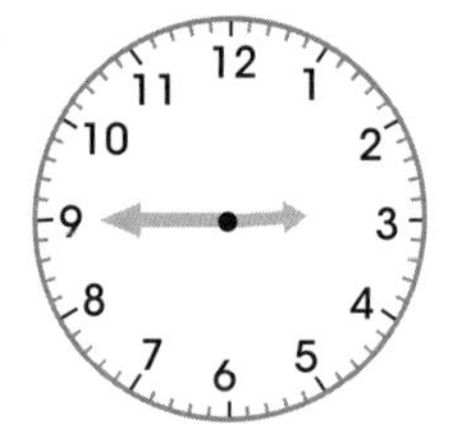

f 

2 Draw the time on each clock face.

a quarter past 4

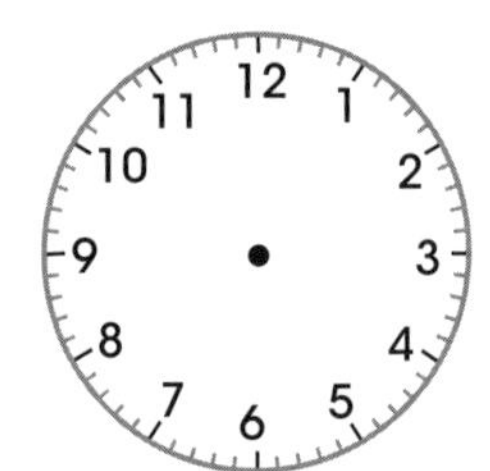

b quarter to 8

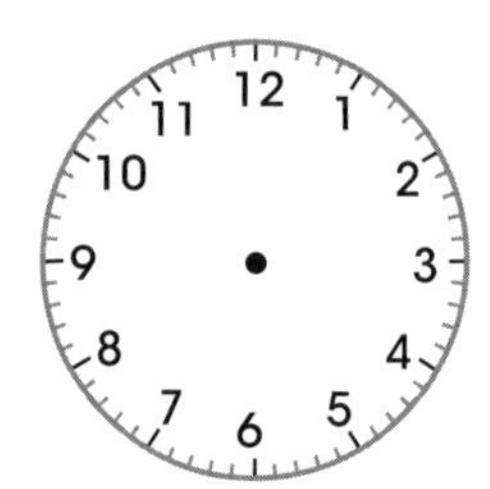

c 10 minutes past 5

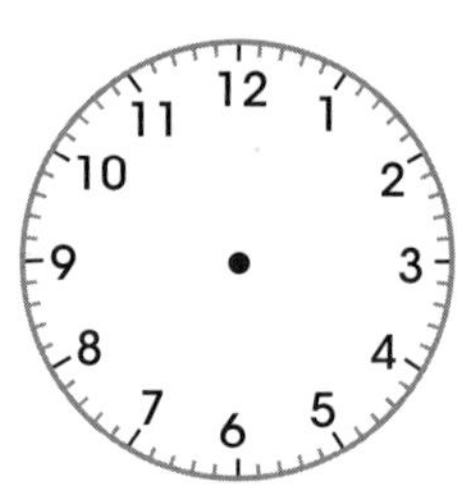

d half past 12

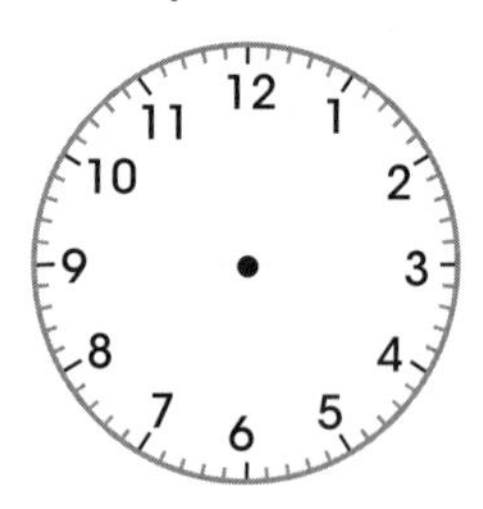

e 11 o'clock

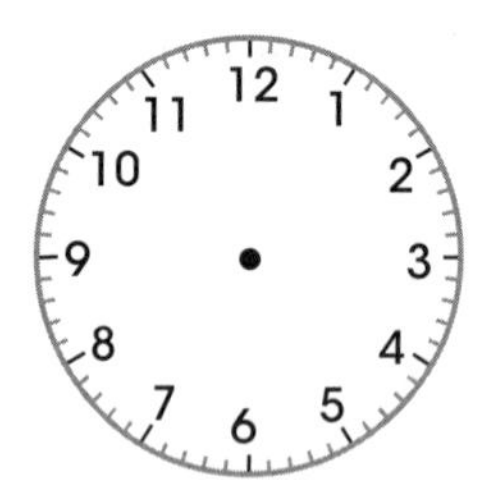

f half past 1

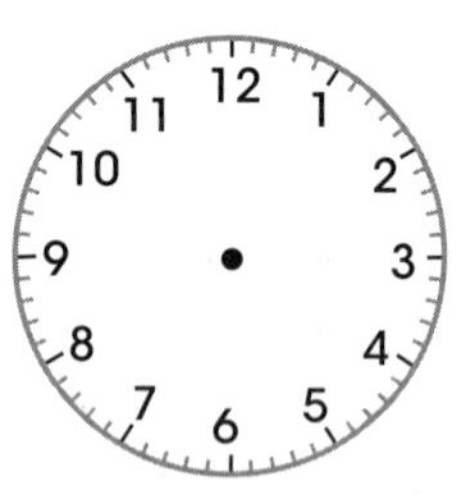

3 What is the difference between half past three and the time shown?

a 

b 

c 

Time (General) (TRB pp. 116–119)
Time MA3-13MG uses 24-hour time and am and pm notation in real-life situations, and constructs timelines

Note: This unit revises content necessary to move on to the specific study of timetables in Unit 26.

Digital and Analogue Time

DATE:

1 Draw the time on each clock face.

a 7:43 pm

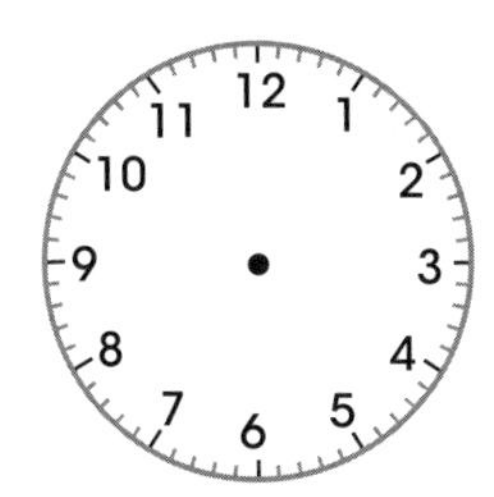

b 9:21 am

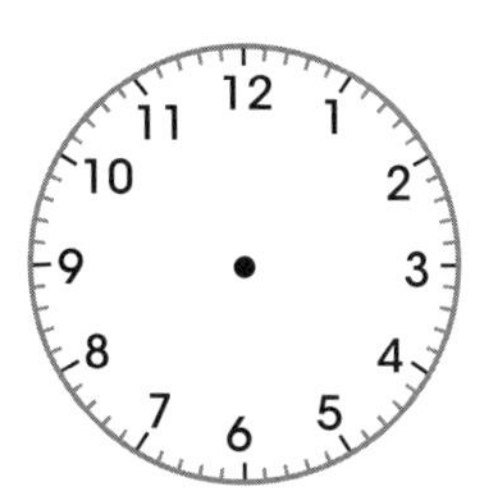

c 4:35 am

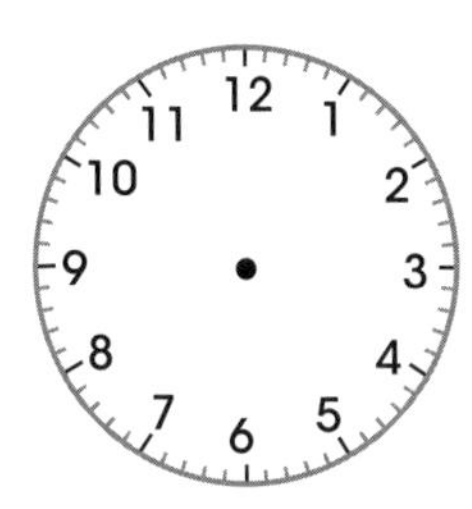

d 11:05 pm

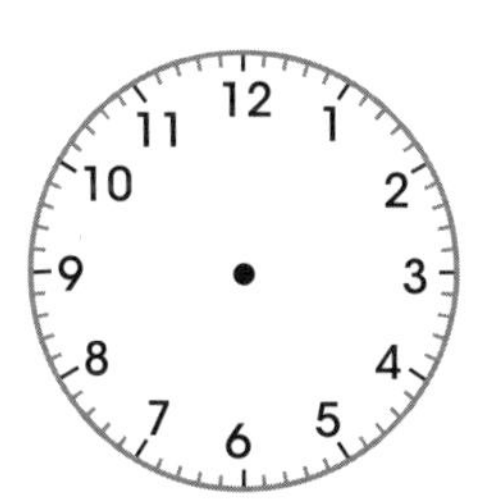

e 2:59 am

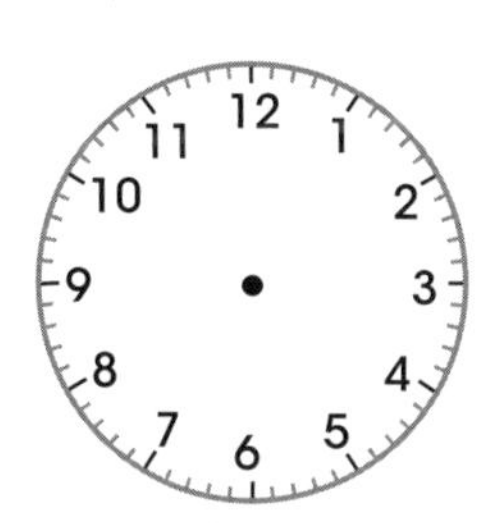

f 6:18 pm

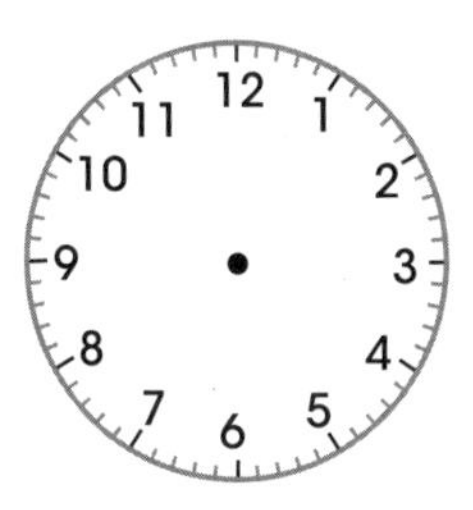

2 Write each time as a digital time.

a quarter to 3 ____________

b quarter past 8 ____________

c 44 minutes past 10 ____________

d 22 minutes past 1 ____________

e 7 minutes to midday ____________

f 12 minutes to 5 ____________

3 Express each time as a digital time.

a

____

b

____

c

____

d

____

e

____

f

____

24-hour Time

DATE:

1 Use 24-hour time to write:

a 8:07 am ______________

b 11:23 am ______________

c 6:15 pm ______________

d 21 to 4 in the afternoon ______________

e 12 past 10 in the morning ______________

f half past 9 in the evening ______________

2 Use am or pm time to write the following as digital times.

a 0120 ______________

b 0714 ______________

c 1700 ______________

d 0352 ______________

e 1243 ______________

f 1427 ______________

3 Find the difference between:

a 5:20 am and 9:15 am ______________

b 6:35 am and 1:15 pm ______________

c 4:35 pm and 8:10 pm ______________

d 9:40 pm and 1:45 am ______________

e 0450 and 1015 ______________

f 0700 and 1525 ______________

4 Dante's flight from Cairns to Melbourne left at 0950 and arrived at 1310. How long was the flight?

Note: This unit revises content necessary to move on to the specific study of timetables in Unit 26.

DATE:

STUDENT ASSESSMENT

1 Complete the table.

	Digital	Analogue	In words	24-hour time
a	3:45 pm			
b				1915
c			Half past 11 at night	

2 Find the difference between the times.

a 0945 and 1420 ________________

b 6:20 am and 1:10 pm ________________

c

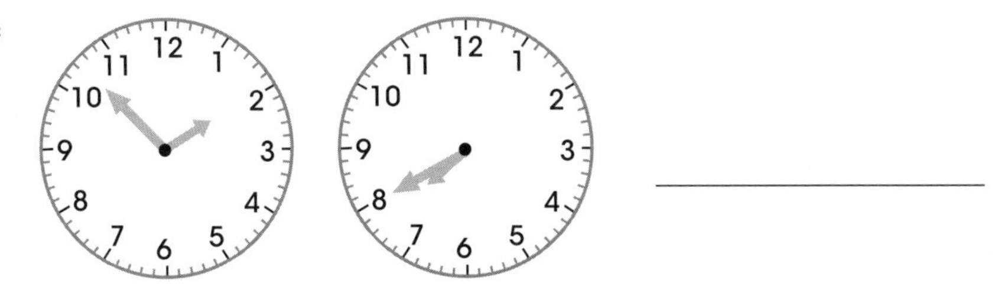

____

d quarter to 7 in the evening and half past 1 in the morning

____

Daily Activities

DATE:

You will need: a partner

This is Toby's timetable for a school week.

Time	Monday	Tuesday	Wednesday	Thursday	Friday
7:15 am	Get up	Get up	Get up	Get up	Get up
8:00 am	Breakfast	Breakfast	Breakfast	Breakfast	Breakfast
9:00 am	Start school	Start school	Start school	Start school	Start school
10:00 am	Maths	Science	Computers	Art	Sport
11:00 am	Morning recess	Morning recess	Morning recess	Morning recess	Morning recess
12:00 pm	Reading	Maths	Reading	Maths	Reading
1:00 pm	Lunch time	Lunch time	Lunch time	Lunch time	Lunch time
2:00 pm	Music	Library	Drama	Reading	Sport
3:30 pm	End school	End school	End school	End school	End school
4:00 pm	Footy training	Mum shopping	Flute	Footy training	Friends
5:00 pm	Footy training	Mum shopping	Flute	Footy training	Friends
6:30 pm	Dinner	Dinner	Dinner	Dinner	Dinner
7:00 pm	TV	Homework	TV	Homework	TV
9:00 pm	Bedtime	Bedtime	Bedtime	Bedtime	Bedtime

1 How much time a week does Toby spend in maths? ______________

2 How much time a week does Toby spend at footy training? ______________

3 How much time a week does Toby spend eating? ______________

4 Write 3 questions of your own about Toby's timetable.

a ______________

b ______________

c ______________

5 Ask your partner for their 3 questions about Toby's timetable. Record the questions and your answers.

a ______________

____

b ______________

____

c ______________

____

Train Timetable

DATE:

This is the train timetable for the Airport Express train in Hong Kong.

The Airport Express runs every 10 minutes.

Station	First train	Last train
Hong Kong	05:50	00:50
Kowloon	05:53	00:53
Tsing Yi	06:00	01:00

Journey	Journey time (mins)
Hong Kong-Airport	24
Kowloon-Airport	21
Tsing Yi-Airport	13

1 How long does the trip take from the airport to the city of Hong Kong? ________

2 If you were at Hong Kong station at 7:45 am, what time would the next train be?

____

3 If you had to get to the airport by 7:00 pm, which train would you need to catch from Kowloon? ________________

4 What is the travel time from Kowloon to Tsing Yi? ________________

5 Write 3 questions for your teacher about the train timetable.

a ____

b ____

c ____

Timelines

DATE:

1 This is a timeline of important events in Faye's life. Take note of the scale.

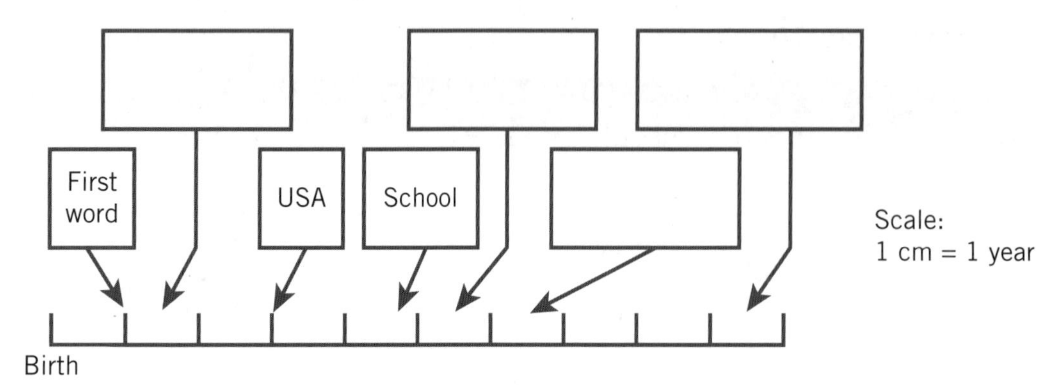

a How old was Faye when she said her first word?

b Faye learned to walk before she went to the USA. Write "Walk" on the timeline.

c Was Faye five years old when she started school? ______________

d Faye was in a school play during her first school year. In the next year, she learned to swim. Write "Play" and "Swim" on the timeline.

e Faye played violin in a concert just after her ninth birthday. Write "Concert" on the timeline and draw an arrow to the correct place.

2 This is a timeline for some special events in 100 years of Australian Rugby League.

Scale: 1 cm = 10 years

1908

Place these events on the timeline by writing the letter for each event.

1947: Parramatta and Manly joined the League. (A)

1976: Easts became the first team to have a sponsor's name on their jerseys. (B)

1908: South Sydney became the first premiers of the League. (C)

1986: Peter Stirling was the first winner of the Clive Churchill medal. (D)

1964: Australia beat France in Australia for the first time. (E)

1919: Australia played its first game in New Zealand. (F)

1957: The first world cup was played in Australia. (G)

2004: A State of Origin game was decided by a golden point for the first time. (H)

1927: New South Wales and Queensland played their first game. (I)

STUDENT ASSESSMENT

1 The hall is used for these sports during the week.

Time (pm)	Monday	Tuesday	Wednesday	Thursday	Friday
4:20–5:00	Netball	Gymnastics	Gymnastics	Gymnastics	Gymnastics
5:00–6:00					
6:00–7:00	Indoor soccer		Indoor soccer	Indoor cricket	
7:00–8:00	Indoor cricket	Netball	Indoor cricket	Netball	Basketball
8:00–9:15	Basketball	Netball	Basketball	Netball	Basketball

a Mia trains for indoor cricket. How much training does Mia have each week? ________________

b How much time is the hall used for on Wednesday? ________________

c What time is the hall closed? ________________

d How long (in minutes) is the first netball session on Thursday? ________

2 Give three different examples of uses of timetables.

____

____

____

3 Make a timeline for one of your typical school days. Include seven events, such as the time you get up, the time you start lessons and the time you have lunch. Draw an arrow to show the time of each event. Make sure you take note of the scale.

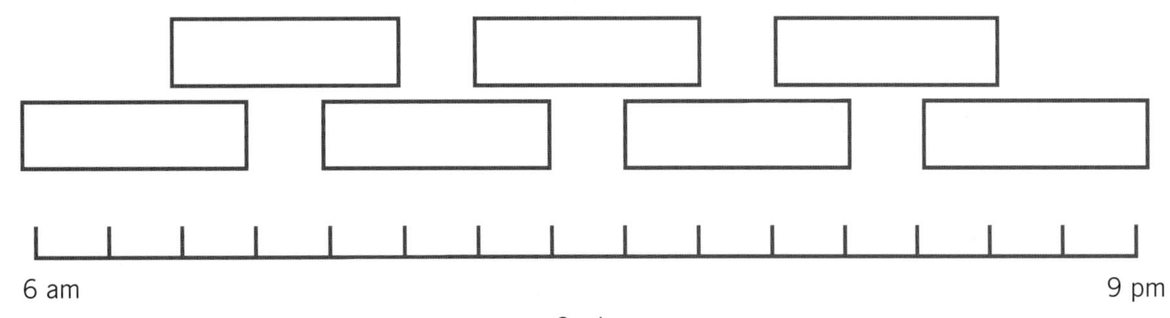

Scale:
1 cm = 1 hour

Fraction of a Group

DATE:

1 Shade the fraction of each group.

a $\frac{1}{4}$ of 8 =

b $\frac{1}{3}$ of 12 =

c $\frac{1}{5}$ of 20 =

d $\frac{1}{8}$ of 16 =

2 Use the array on the right to find:

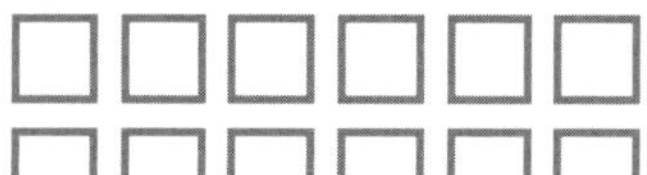

a $\frac{1}{2}$ of 24 =

b $\frac{1}{6}$ of 24 =

c $\frac{2}{3}$ of 24 =

d $\frac{3}{4}$ of 24 =

3 Draw a picture to represent the following.

a $\frac{1}{4}$ of 16 =

b $\frac{2}{5}$ of 10 =

c $\frac{2}{3}$ of 9 =

d $\frac{4}{10}$ of 20 =

Fraction of a Quantity

DATE:

1 Find the following.

a $\frac{1}{5}$ of 20 =

b $\frac{1}{3}$ of 15 =

c $\frac{1}{8}$ of 24 =

d $\frac{1}{4}$ of 32 =

2 Find the following.

a $\frac{2}{3}$ of 18 =

b $\frac{3}{5}$ of 20 =

c $\frac{3}{8}$ of 16 =

d $\frac{6}{10}$ of 60 =

3 Answer the problems.

a Abdul had 24 footy cards and traded $\frac{3}{8}$ of them.
How many cards did he trade?

b Riley had 27 emails. He replied to $\frac{2}{3}$ of them.
How many emails did he reply to?

c Amity had a packet of 20 stamps. She used $\frac{3}{10}$ of them.
How many stamps does she have left?

d Sienna had 30 plants to put in pots. She planted $\frac{5}{6}$ of them.
How many does she have left to plant?

e Lee had to bake 40 loaves of bread. He made $\frac{3}{5}$ of them.
How many loaves does he still have to bake?

f Yuko has 48 books on her bookshelf. She has read $\frac{3}{4}$ of them.
How many books does she have left to read?

Creating Fractions

DATE:

On one shelf in a shoe store, there are 24 pairs of shoes in these colours:
red, green, blue, yellow, orange, pink, red, purple, purple, brown, green, blue, yellow, orange, red, brown, green, blue, purple, red, yellow, blue, blue, green.

1 Using the data above, complete the table. One row has been done.

Colour	Tally	Total	As a fraction	Simplified
Red	IIII	4	$\frac{4}{24}$	$\frac{1}{6}$
Green				
Blue				
Yellow				
Orange				
Pink				
Purple				
Brown				

2 Total the fractions in the "As a fraction" column. What do you find?

____

3 Write three questions about the fractions and their relationships to the colours. For example, "What colour is the same fraction of the total as yellow?"

____

____

____

Extension: Swap your questions with a partner. Write their questions and your answers.

____

____

____

DATE:

STUDENT ASSESSMENT

1 Use the array on the right to find:

a $\frac{1}{20}$ of 20 =

b $\frac{3}{10}$ of 20 =

c $\frac{7}{20}$ of 20 =

d $\frac{3}{4}$ of 20 =

2 Draw a picture to represent $\frac{2}{3}$ of 15.

3 Find the following.

a $\frac{2}{3}$ of 9 =

b $\frac{3}{4}$ of 16 =

c $\frac{3}{10}$ of 40 =

d $\frac{1}{9}$ of 18 =

4 Cooper kicked $\frac{5}{6}$ of 12 goals scored by his team. How many goals did he kick?

5 Cars of the following colours drove down the street.

a Express each colour as a fraction of the total.

Colours	Cars	Fraction of the total
Red	9	
Blue	7	
Green	13	
Yellow	5	
White	16	

b Write 2 statements about the car colours.

____

____

Decimals and Fractions

DATE:

1 Complete the table.

	Diagram	Fraction	Decimal		Diagram	Fraction	Decimal
a	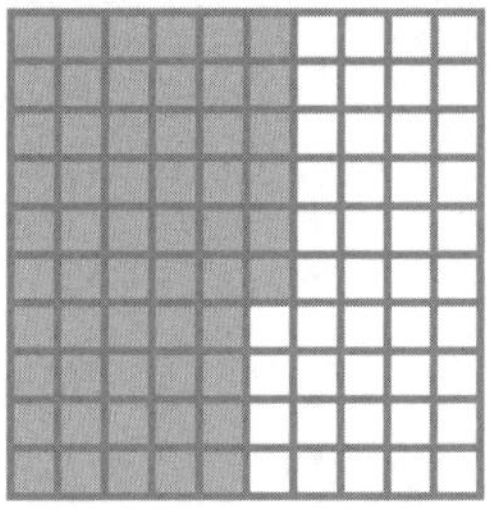			**b**	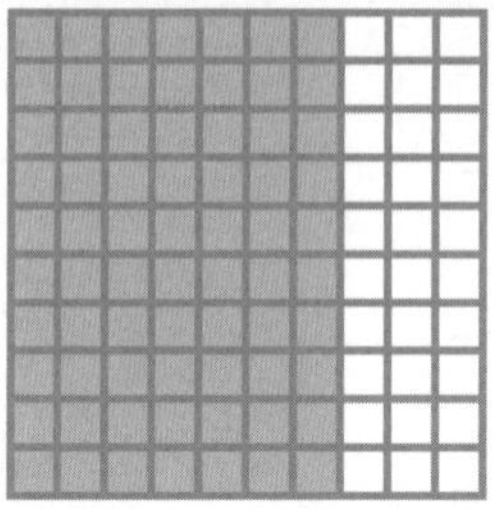		
c	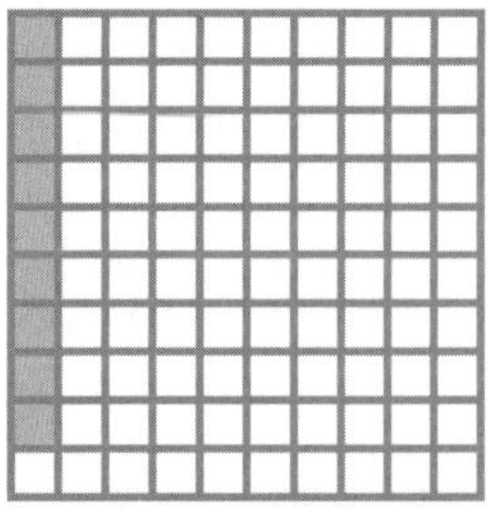			**d**	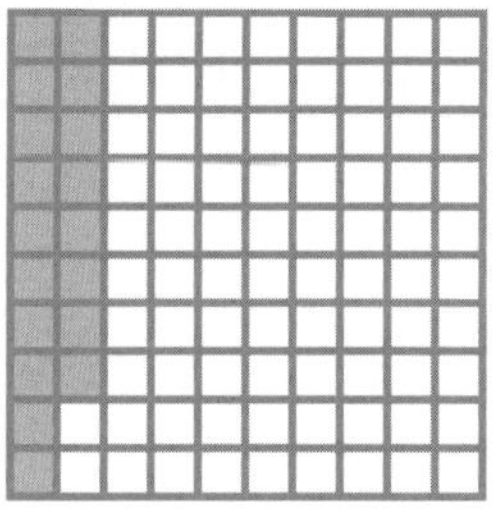		

2 Write the fraction for each decimal.

a 6 tenths =
b 9 tenths =
c 15 hundredths =

d 29 hundredths =
e 37 hundredths =
f 80 hundredths =

3 Write the decimal for each fraction.

a $\frac{3}{10}$ =
b $\frac{7}{10}$ =
c $\frac{25}{100}$ =

d $\frac{37}{100}$ =
e $\frac{90}{100}$ =
f $\frac{3}{100}$ =

4 Express the following as fractions.

a 0.2 =
b 0.08 =
c 0.14 =

d 0.47 =
e 0.98 =
f 0.40 =

Extension: Express the following as fractions.

a 1.25 =
b 3.02 =

c 9.1 =
d 4.38 =

Decimals and Percentages

DATE:

1 Complete the table.

	Diagram	Percentage	Decimal		Diagram	Percentage	Decimal
a				**b**			
c				**d**			

2 Express each percentage as a decimal.

a 41% =

b 95% =

c 5% =

d 30% =

e 16% =

f 125% =

3 Express each decimal as a percentage.

a 0.2 =

b 0.9 =

c 0.09 =

d 0.36 =

e 0.87 =

f 0.11 =

Extension: Find:

a 10% of $50 ___________

b 25% of $40 ___________

c 30% of $60 ___________

d 50% of $82 ___________

Percentages, Decimals and Fractions

DATE:

1 Complete the table.

	Diagram	Fraction	Percentage	Decimal
a	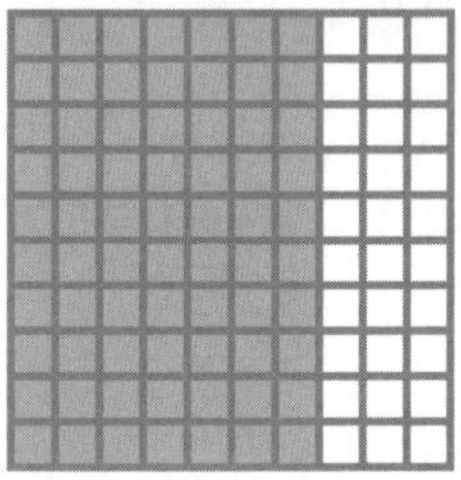			
b	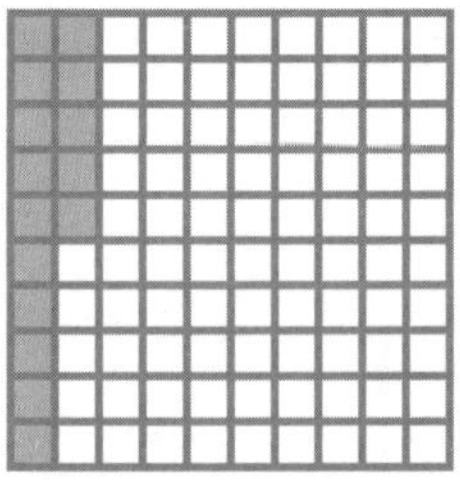			
c	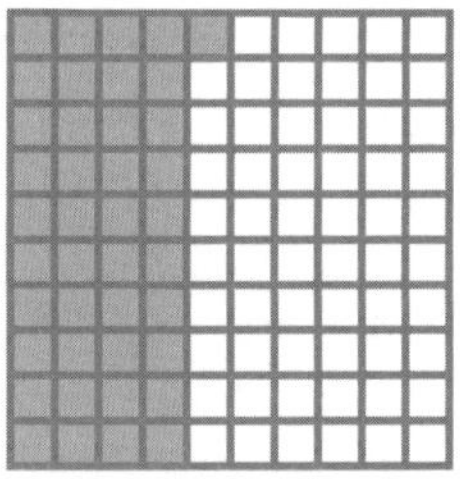			
d	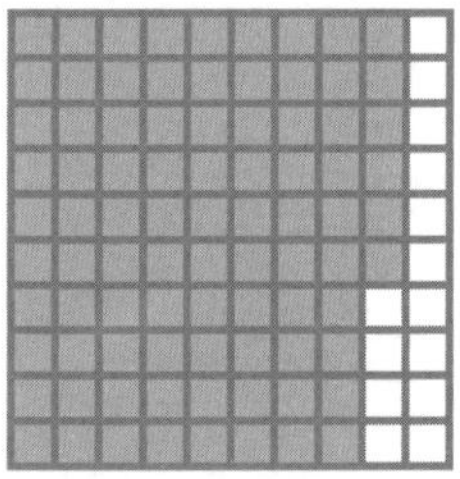			

2 Express each fraction as a percentage.

a $\frac{80}{100} =$

b $\frac{7}{100} =$

c $\frac{47}{100} =$

d $\frac{21}{100} =$

3 Circle the largest value in each group.

a 0.4, $\frac{4}{100}$, 44%

b 99%, $\frac{98}{100}$, 0.9

c 0.33, $\frac{3}{10}$, 37%

d 1.5, $\frac{156}{100}$, 152%

DATE:

STUDENT ASSESSMENT

1 Express the following as decimals.

a 63% =

b 40% =

c $\frac{5}{100}$ =

d $\frac{3}{100}$ =

e $\frac{21}{100}$ =

f $\frac{60}{100}$ =

2 Express the following as fractions.

a 90% =

b 7% =

c 14% =

d 0.63 =

e 0.30 =

f 0.1 =

3 Express the following as percentages.

a $\frac{51}{100}$ =

b $\frac{9}{100}$ =

c $\frac{4}{10}$ =

d $\frac{50}{100}$ =

e 0.20 =

f 0.58 =

4 Find the following.

a 10% of 40 elephants = ______________________

b 30% of 60 flamingoes = ______________________

c 20% of 100 tarantulas = ______________________

d 5% of 20 goats = ______________________

5

	Diagram	Percentage	Decimal		Diagram	Percentage	Decimal
a	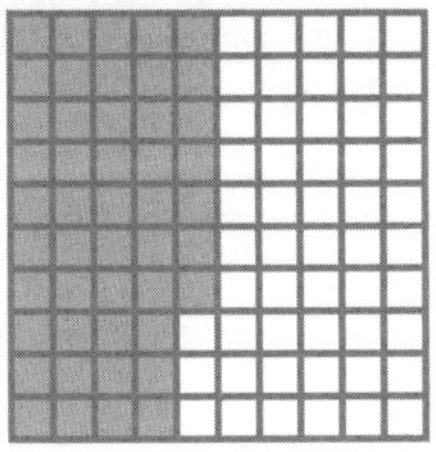			**b**	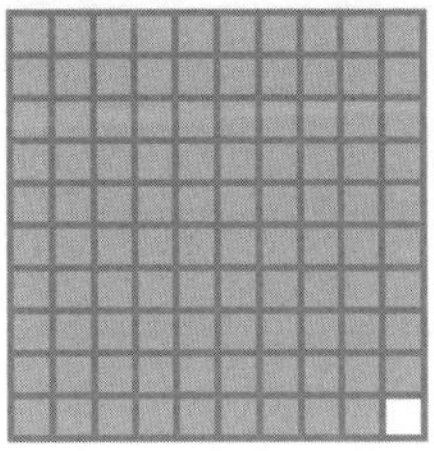		

Collecting My Data: Sustainability

DATE:

You will need: a partner

1 My partner: ______________________

2 Our topic: ______________________

3 Data we have collected: ______________________

4 What we found (3 comments):

Displaying Data

DATE:

1 In pairs, create a survey and collect data about musical instruments played by people in your class. Include aspects such as if the participant is female or male, what instrument they play, and how often they practise.

---

2 Using the data above, create a graph.

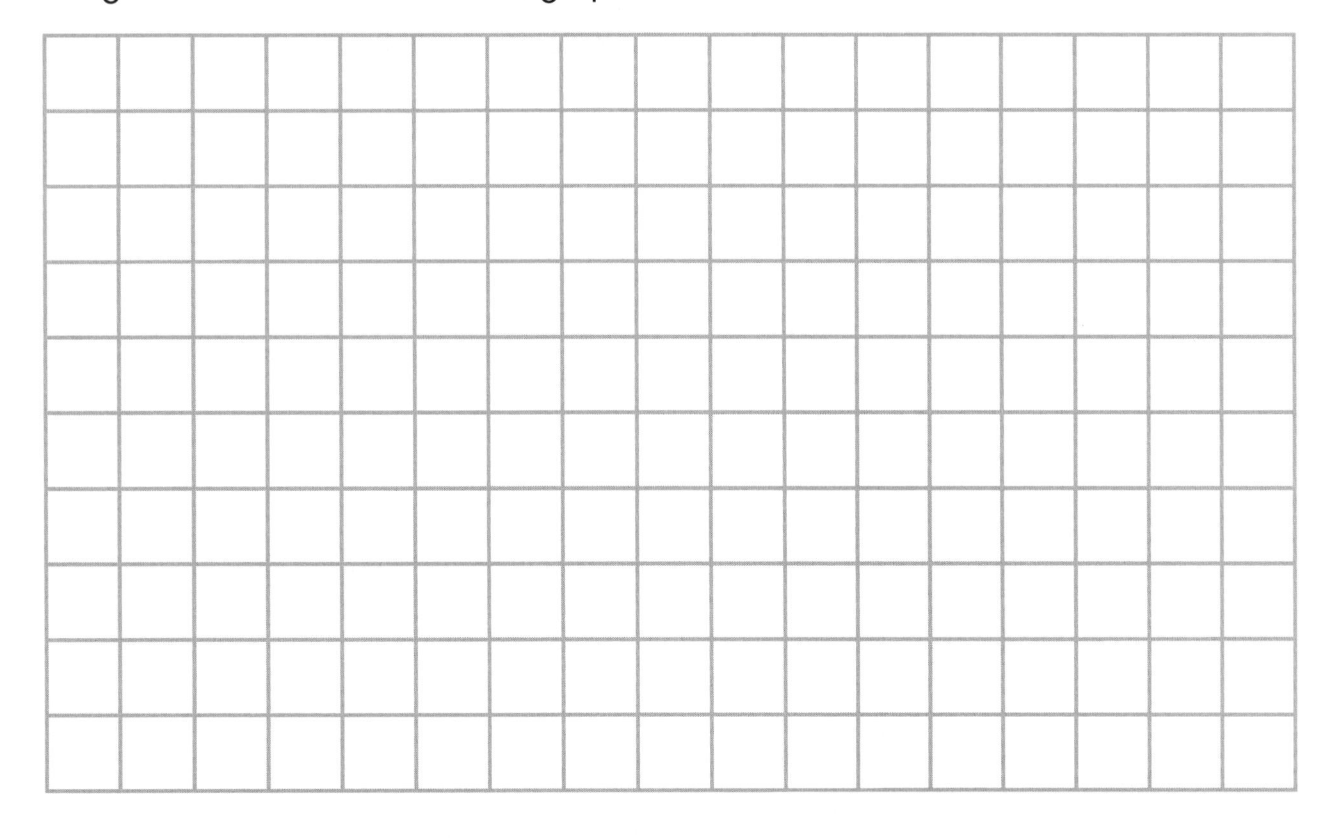

3 Write 3 comments about what you found.

Finding Data

DATE:

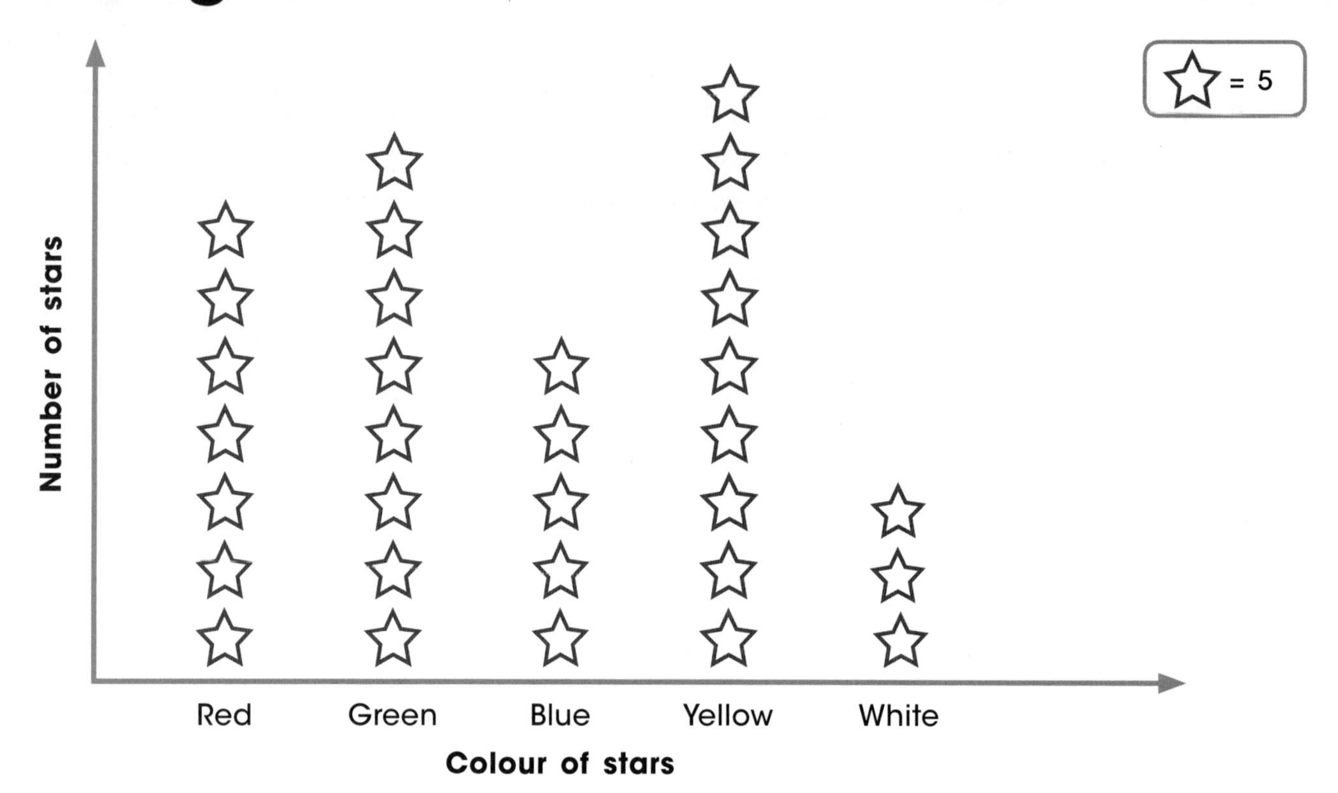

1 Write 3 questions about the graph.

____

____

____

2 From the graph, create a table of data.

3 Give 3 examples of what the data could be representing. For example, stars collected for school house points.

____

____

____

STUDENT ASSESSMENT

1 List 3 different ways of collecting data.

____

____

____

2 Describe all of the different features you can find on a graph. (You may wish to draw an example and label it.)

____

____

____

____

____

____

3 Create a table of information about your favourite animals. Include at least 5 different animals.

4 Sketch a graph of your data from Question 3.

Collecting and Interpreting Data

DATE:

You will need: a partner

1 My partner: ______________________________

2 Our topic: ______________________________

3 How we decided on the topic: ______________________________

4 Any notes about our information:

____

____

____

____

____

____

____

5 What we found (include 5 interpretative statements based on your analysis of the data):

____

____

____

____

____

6 As a result of our investigation, what we would investigate next is:

____

____

____

____

____

____

____

Media Example

DATE:

You are a writer for the local newspaper. You are to write a short article about local traffic, including the use of data.

Note: This activity could be factual or fictional (check with your teacher).

Misleading Data

DATE:

Here is some data about maths tests.

Student	Test 1 (out of 20)	Test 2 (out of 20)	Test 3 (out of 20)	Test 4 (out of 20)
Tarita	15	20	16	19
Gabriel	16	18	17	18
Aiden	17	17	17	18
Lily	19	19	18	19
Yan	20	20	18	20
Elsa	10	15	16	14
Min-jun	15	9	11	13
Harry	18	20	19	20
Sarah	17	16	18	18
Mariam	16	17	17	18
Ali	15	15	14	16
Charlotte	11	13	12	10

1 Your task is to create a graph that misleads a reader about the student test results.

2 Describe how your graph is misleading.

STUDENT ASSESSMENT

This is a section of a graph about rainfall.

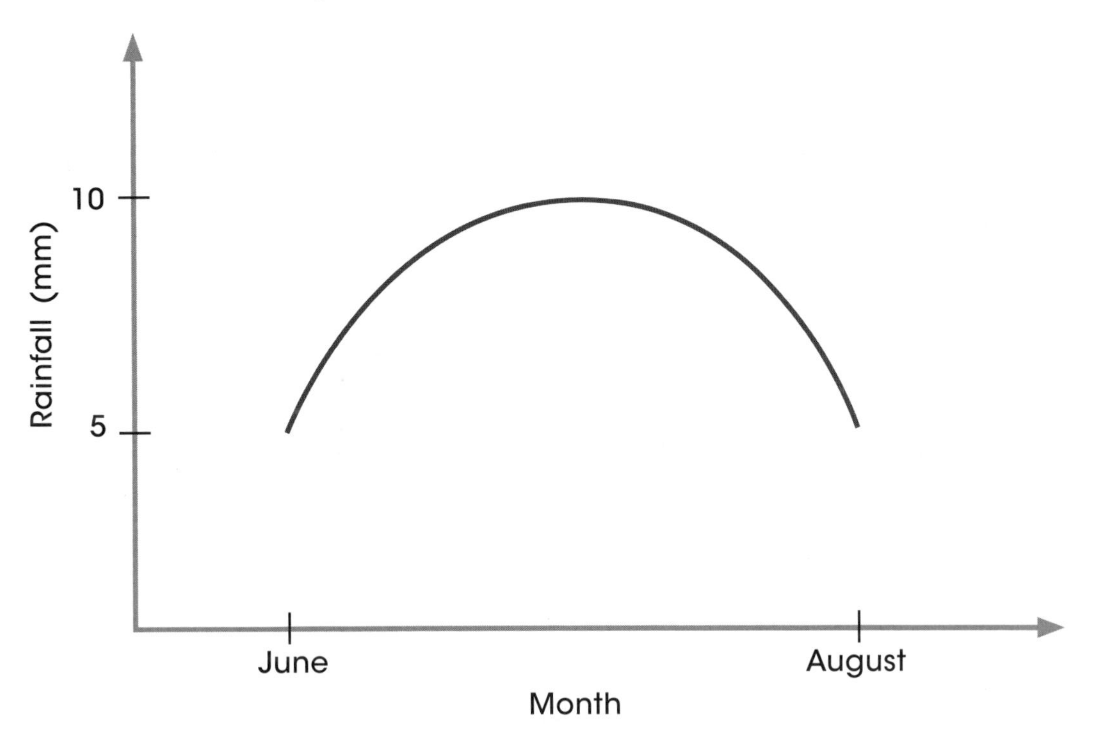

1 Write 2 statements about the graph.

____

____

____

____

2 Describe how the representation could be misleading.

____

____

____

____

3 When might this type of data be used in the media?

____

____

____

Finding the Percentage

DATE:

You may need: a calculator

1 State the percentage shaded for each hundreds square.

a

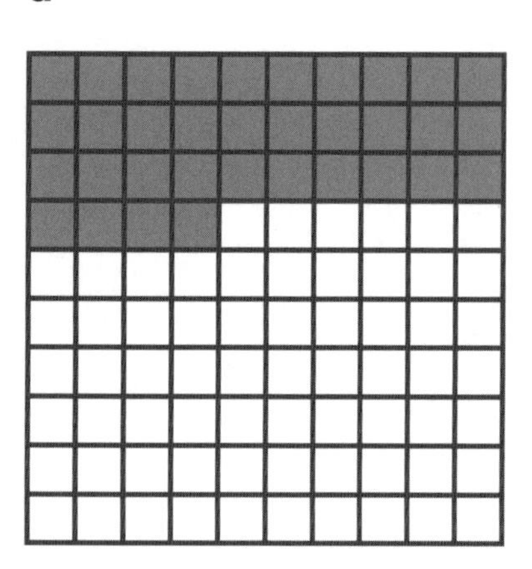

b

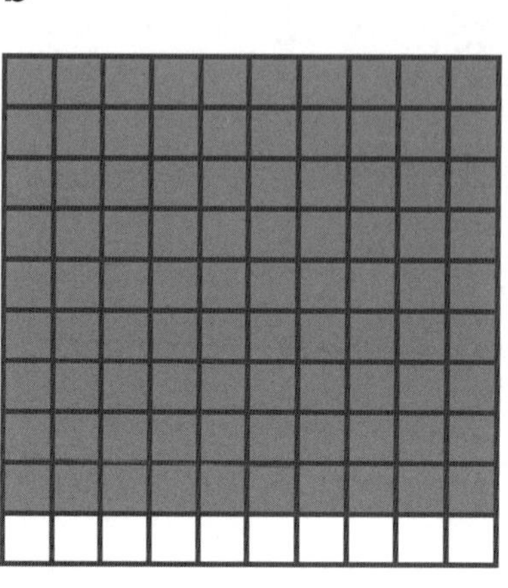

c

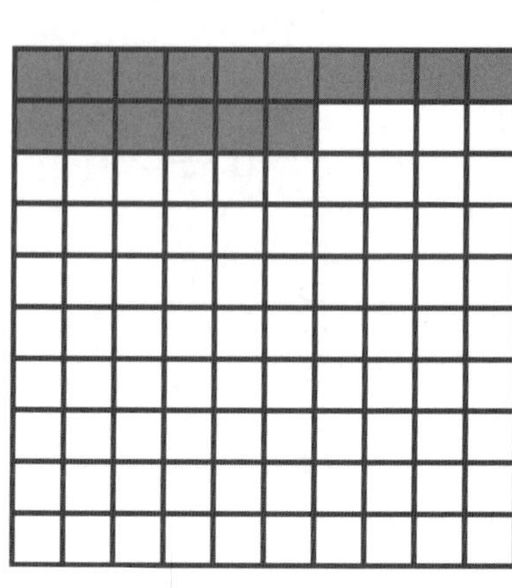

2 Complete the table.

	Percentage shaded	Percentage un-shaded
a	43%	
b		82%
c	11%	
d		27%

3 Find:

a 20% of $1 ____________ **b** 95% of $1 ____________

c 15% of $1 ____________ **d** 69% of $1 ____________

4 Find:

a 25% of $40 ____________ **b** 10% of $60 ____________

c 5% of $100 ____________ **d** 20% of $30 ____________

5 Find:

a 20% of 800 ____________ **b** 50% of 900 ____________

c 10% of 750 ____________ **d** 25% of 120 ____________

Percentage Discount

DATE:

You may need: a calculator

1 Find each amount.

a 50% of $90 __________

b 20% of $50 __________

c 15% of $110 __________

d 25% of $45 __________

2 If you took away, or discounted, the following amounts, how much money would be left?

	Amount taken	Amount left
a 10% of $42		
b 25% of $88		
c 20% of $64		
d 50% of $75		

3 Complete the table.

	Original price	Discount as a percentage	Discount as an amount	New price
a	$500	20%		
b	$700	25%		
c	$600	30%		
d	$200	10%		

4 Find the sale price of:

a a $125 DVD player with 25% off __________

b a $40 book with a 20% discount __________

c a $300 MP3 player with 10% off __________

Extension: If a bag on sale costs $40 after being reduced by 20%, what was the original price of the bag? __________

Value for Money

DATE:

You may need: a calculator

A shop sells items in 3 different ways. You can buy items individually, you can buy them in bulk, or you can buy them at a higher price but with a percentage discount. Look at the sets of items and decide in each case which way is cheaper, per item.

	Cost of 1 item	Cost of bulk items	% off higher price	Which of the 3 options is cheaper?
a	40c	10 for $4.90	50c each with 10% off	
b	60c	20 for $11.00	90c each with 30% off	
c	45c	8 for $3.68	50c each with 20% off	
d	95c	10 for $9.00	$1.00 each with 10% off	
e	$1.10	6 for $6.30	$1.50 each with 40% off	
f	$5.90	8 for $48.00	$6.00 each with 5% off	
g	$11.00	10 for $100.00	$20.00 each with 50% off	

DATE:

STUDENT ASSESSMENT

You may need: a calculator

1 If 20% of a hundreds square is shaded, what percentage is un-shaded? ___________

2 Find:

a 20% of $10 ___________

b 50% of $16 ___________

c 10% of $2 ___________

d 25% of $30 ___________

3 Find:

a 25% of 400 drummers ___________

b 20% of 200 dancers ___________

c 10% of 150 teachers ___________

d 30% of 500 students ___________

4 If a $95 computer game is reduced by 20%, what is the sale price of the game?

5 If a $36 T-shirt is reduced by 30%, what is the sale price of the T-shirt?

6 Which is cheaper per item: a pen for $1.30, 10 pens for $12.20 or a pen for $1.50 with 20% off?

7 Explain what percentage means.

____

____

Translations and Reflections

DATE:

1 Describe each term.

a reflection: ___

b translation: ___

2 Translate each shape.

a

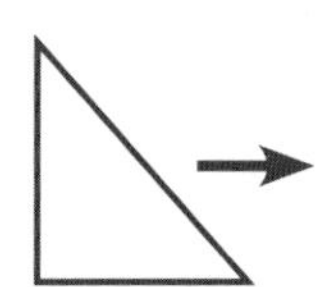

b

c

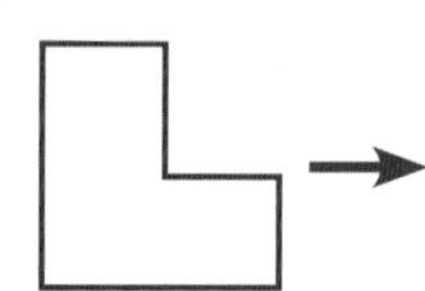

3 Reflect each shape.

a

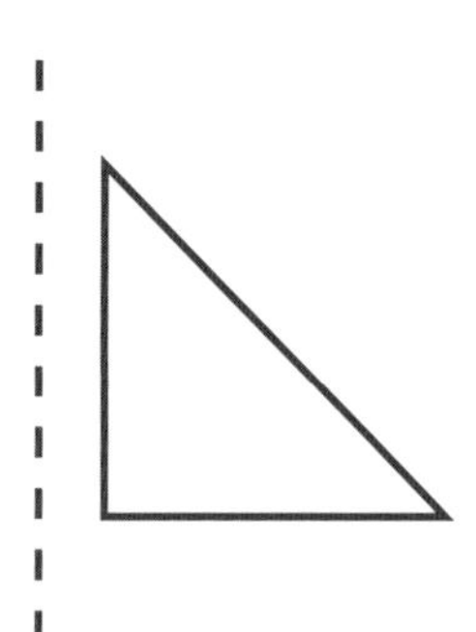

b

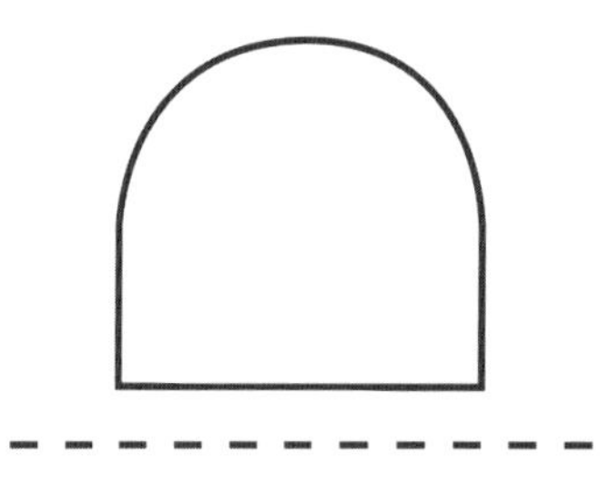

c

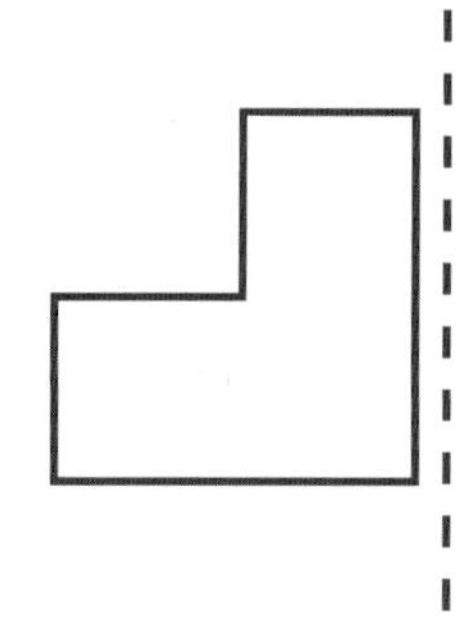

4 Draw the following reflections and translations in the grid.

a Reflect the triangle to the right.

b Reflect the new triangle down.

c Translate the new triangle to the right 2 units.

d Reflect the new triangle down.

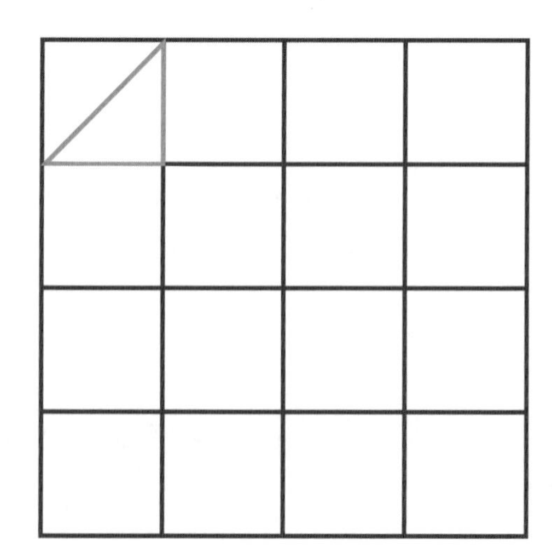

Extension: On another sheet of paper, create your own puzzle based around translations and reflections.

Rotations

DATE:

1 Rotate each shape to the right.

a 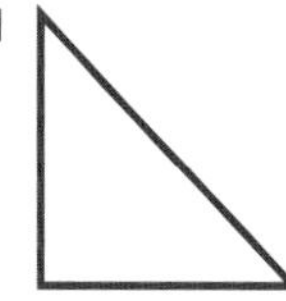

b 

c 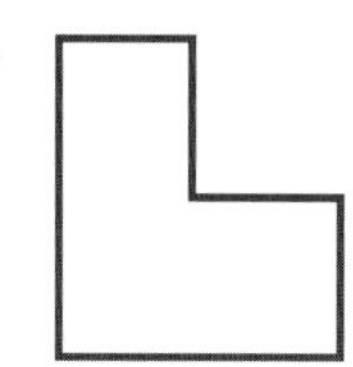

2 Rotate each shape to the left.

a 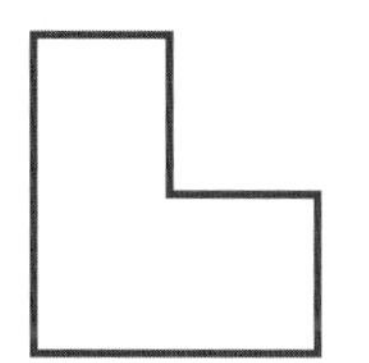

b 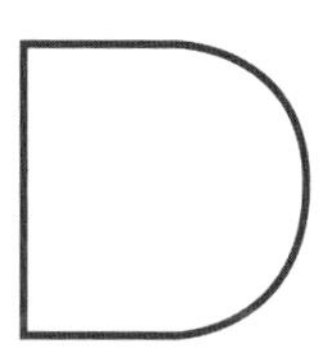

c 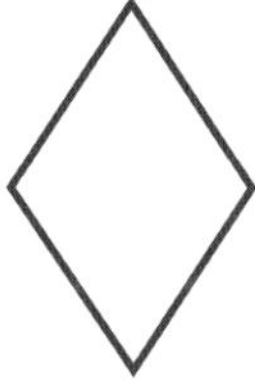

3 Rotate the book across the bookshelf.

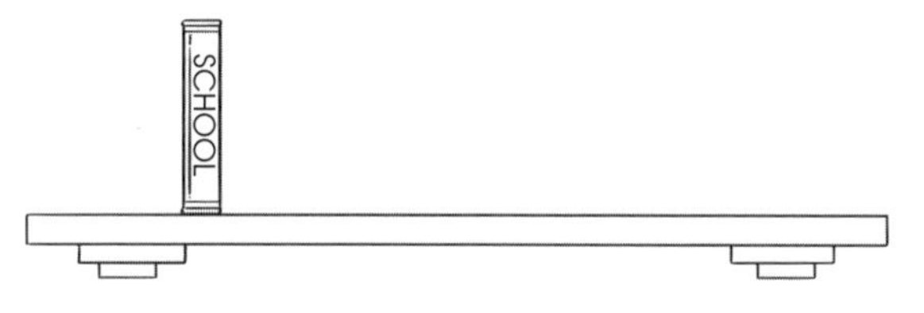

4 Rotate the box across the floor.

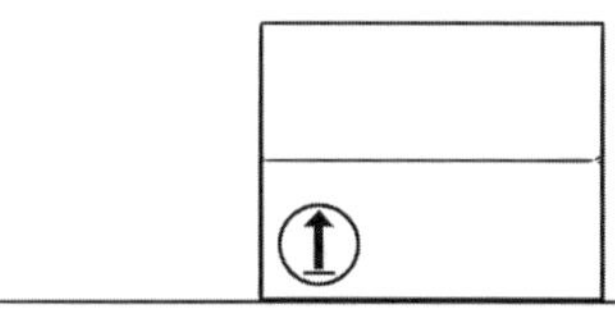

Extension: Design your own pattern based on rotation.

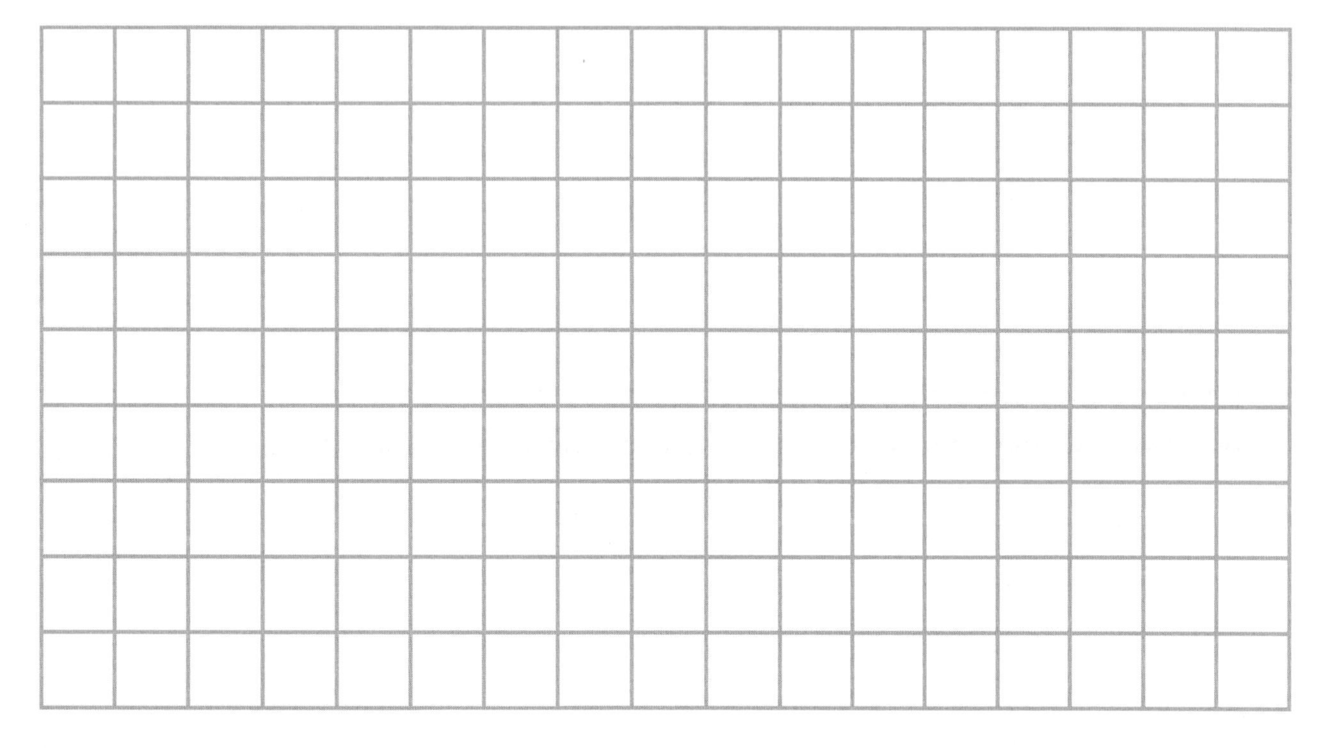

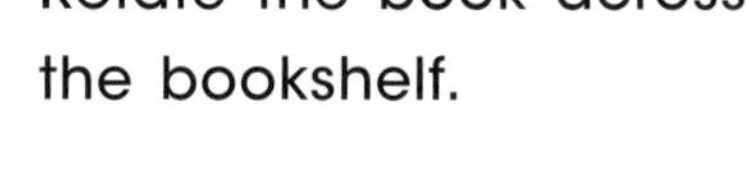

Swimwear Design

DATE:

Your task is to design a pair of bathers or board shorts using reflection, rotation and translation of shapes.

1 Create your design here.

2 Describe how you used the different transformations in your design.

DATE:

STUDENT ASSESSMENT

1 Complete the table.

Transformation	Definition	Example diagram
Reflection		
Rotation		
Translation		

2 Give three examples of when transformations might be used.

____

____

____

3 Draw a simple design based on transformations. Include at least one example each of reflection, rotation and translation.

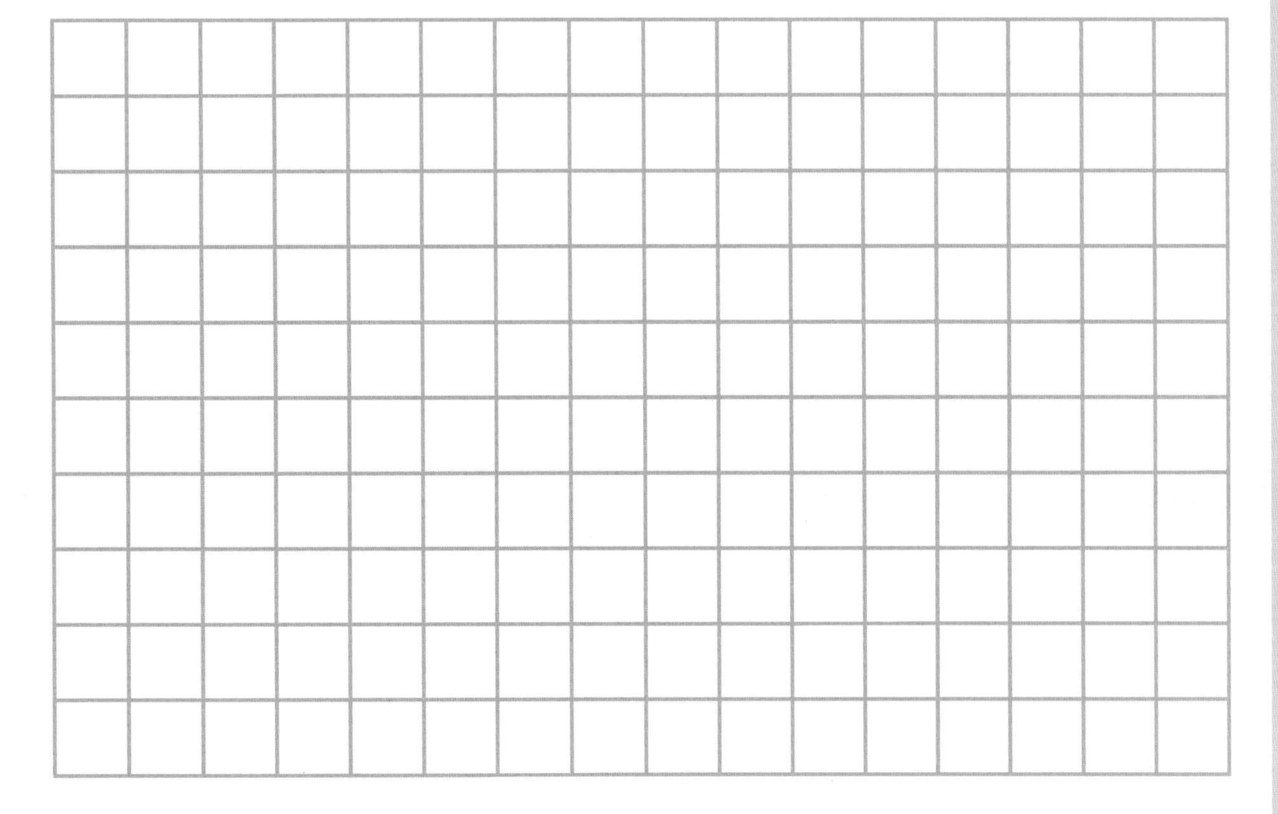

Finding Angles

DATE:

1 Name each type of angle.

a

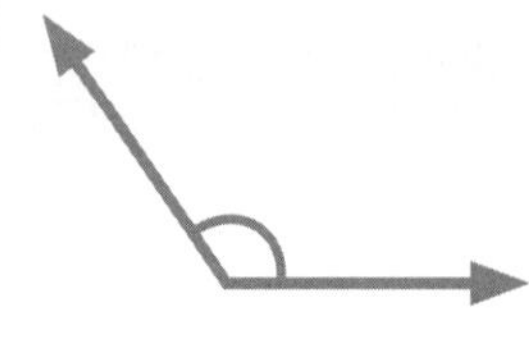

____

b

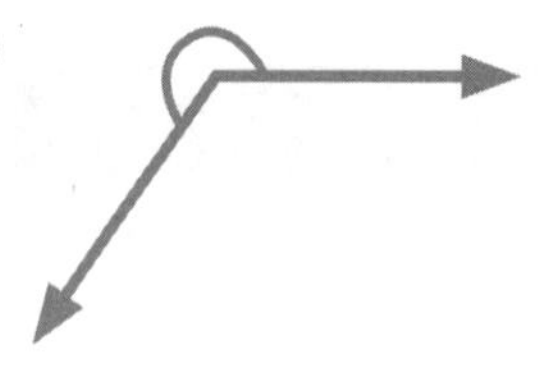

____

c

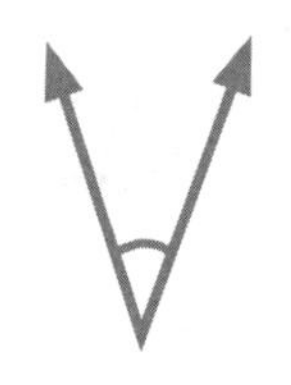

____

2 Estimate the size of each angle.

a

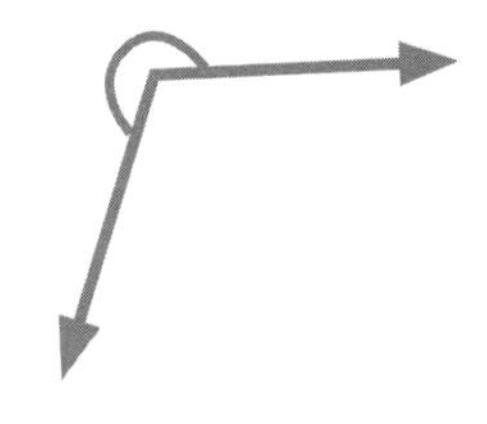

____

b

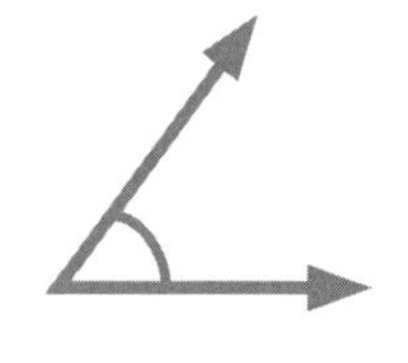

____

c

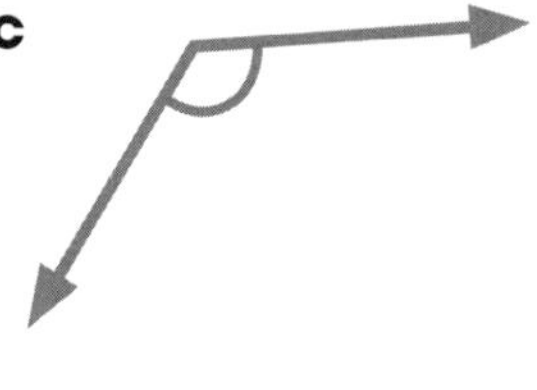

____

3 Find the value of each letter, without using a protractor.

a

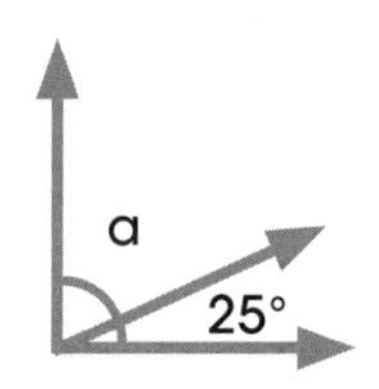

a = __________

b

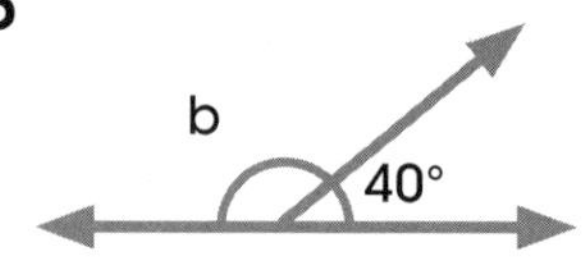

b = __________

c

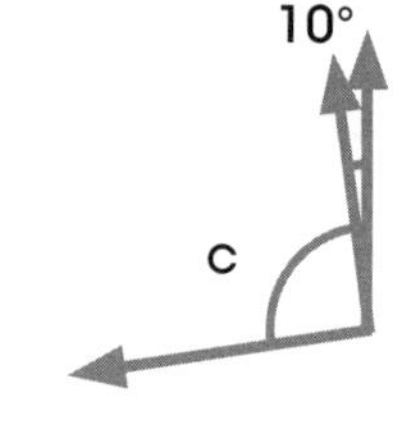

c = __________

d

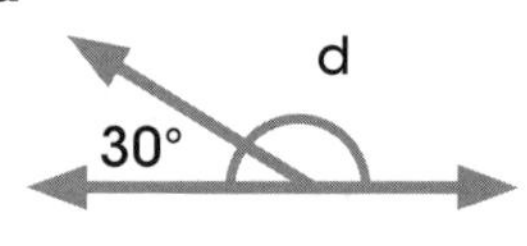

d = __________

e

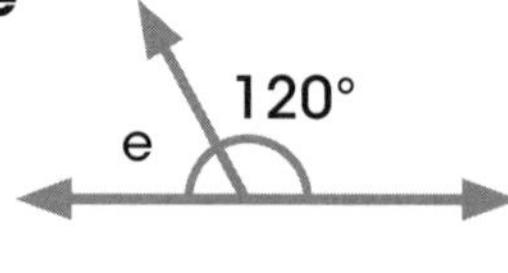

e = __________

f

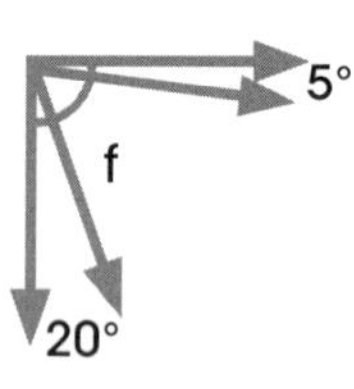

f = __________

Extension: This is the top of a roof.
What is the missing angle x?

x = __________

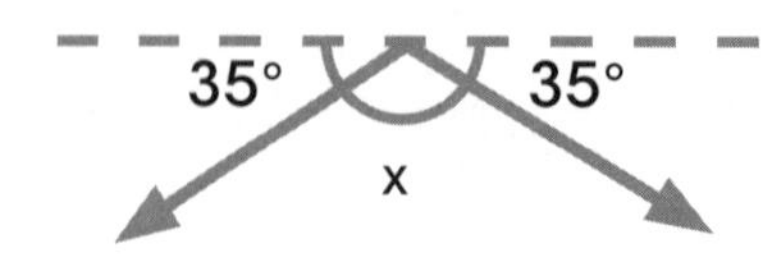

Vertically Opposite Angles

DATE:

1 Find the value of each angle.

a

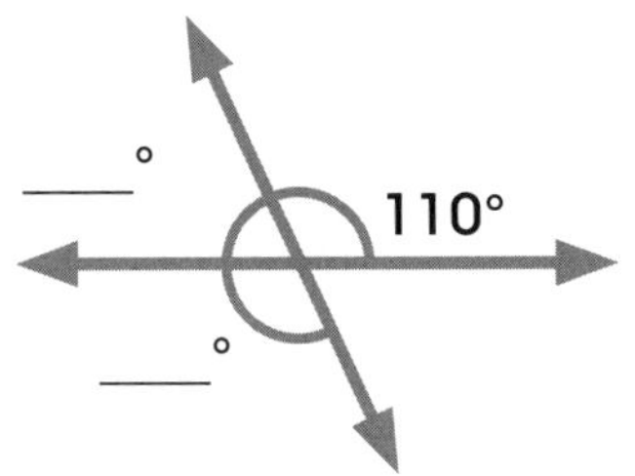

b

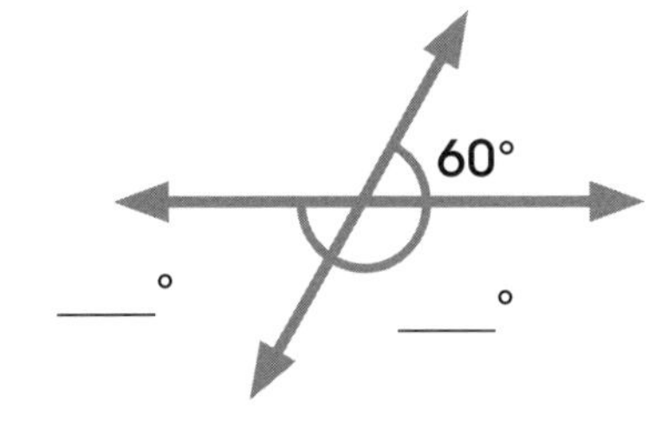

c

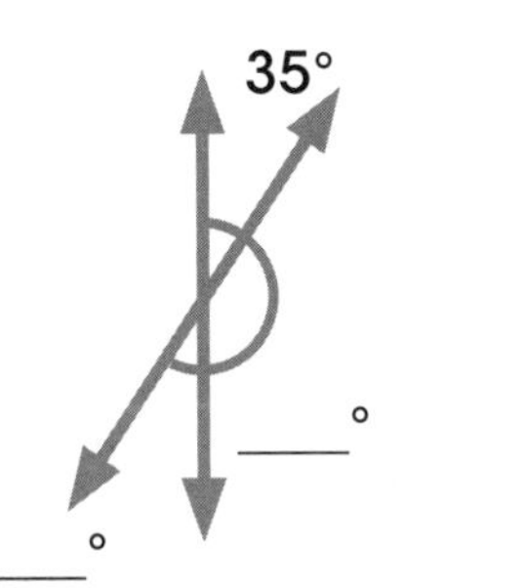

d

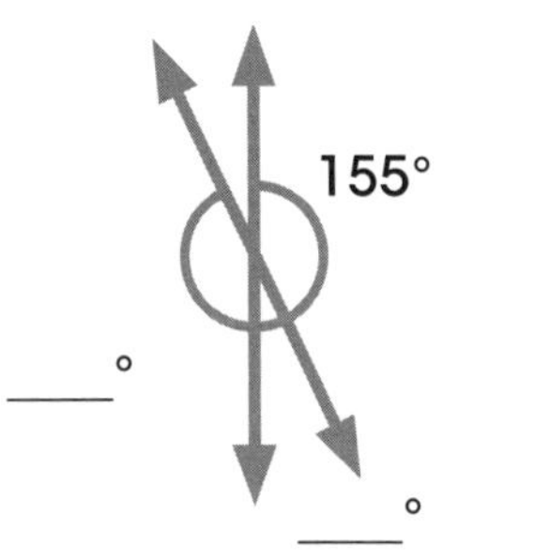

2 Find the value of each angle.

a

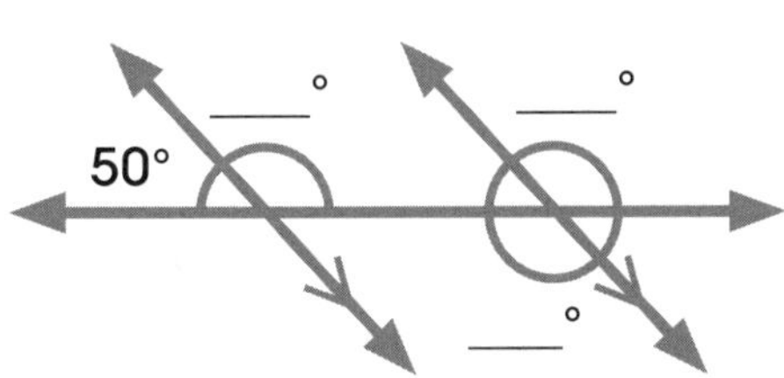

b

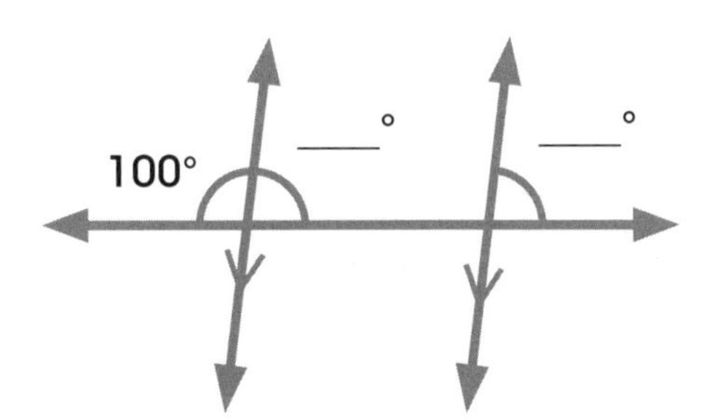

c

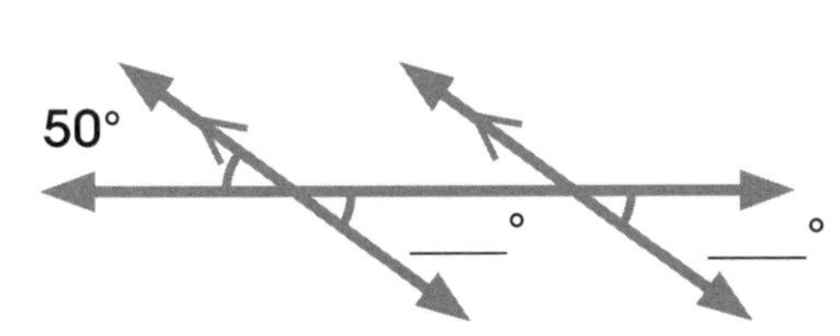

d

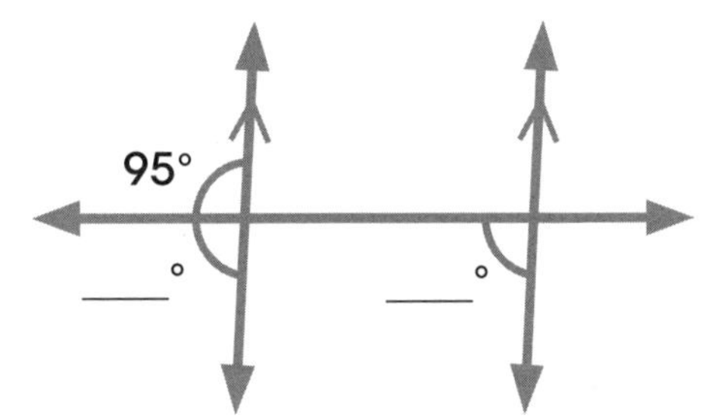

3 Find the value of each angle.

a ____°

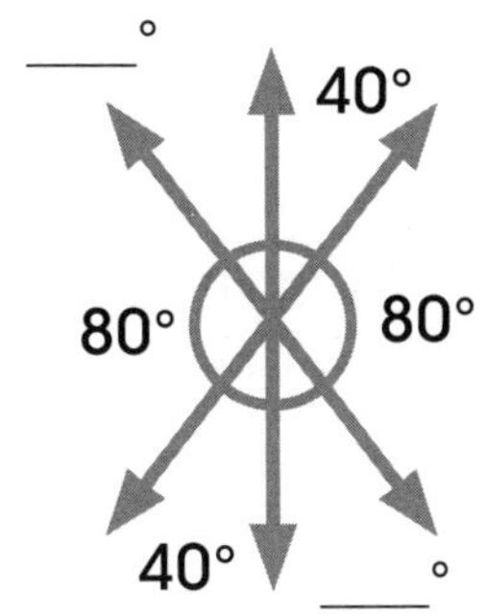

b

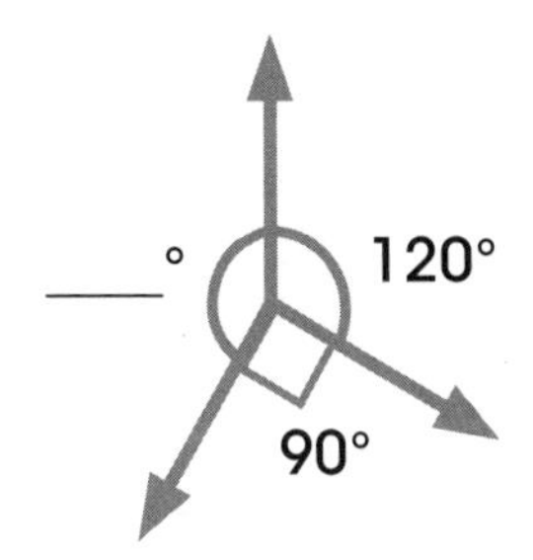

Symmetry

DATE:

1 Draw the lines of symmetry for each shape.

a

b

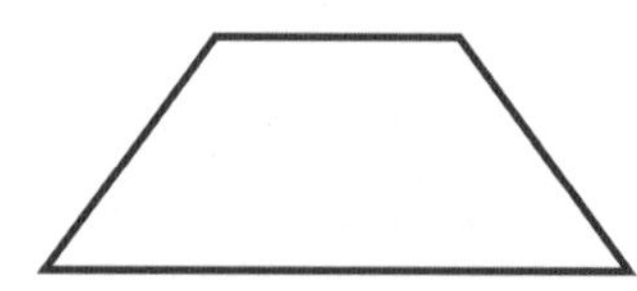

c

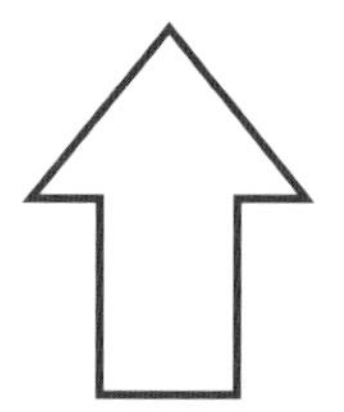

d

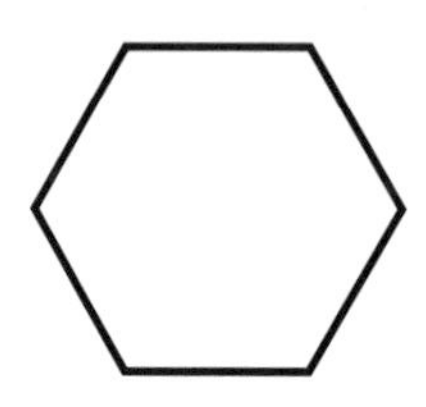

2 Draw the lines of symmetry in each image.

a

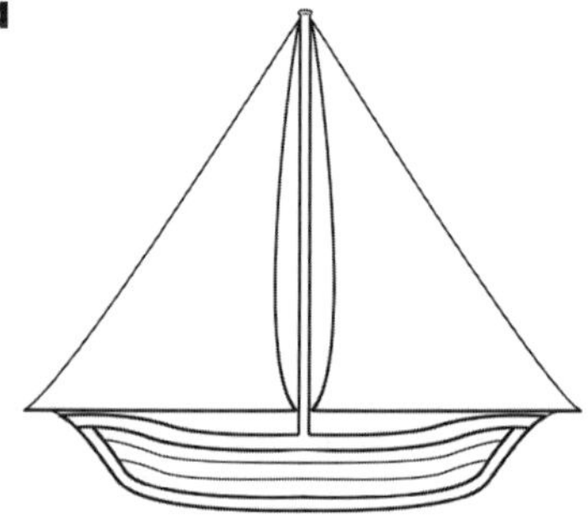

b

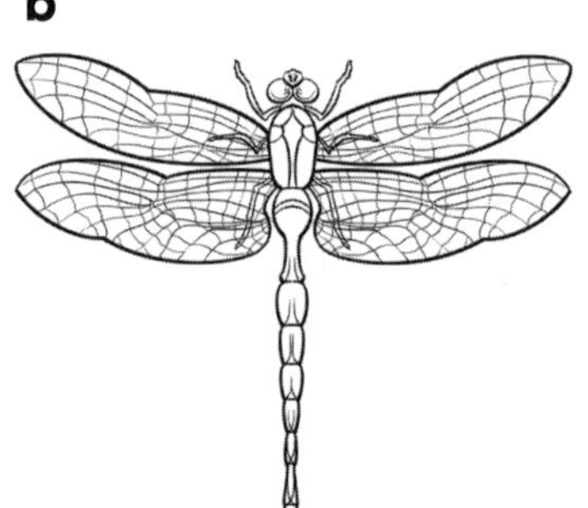

c

3 Complete each image to make it symmetrical.

a

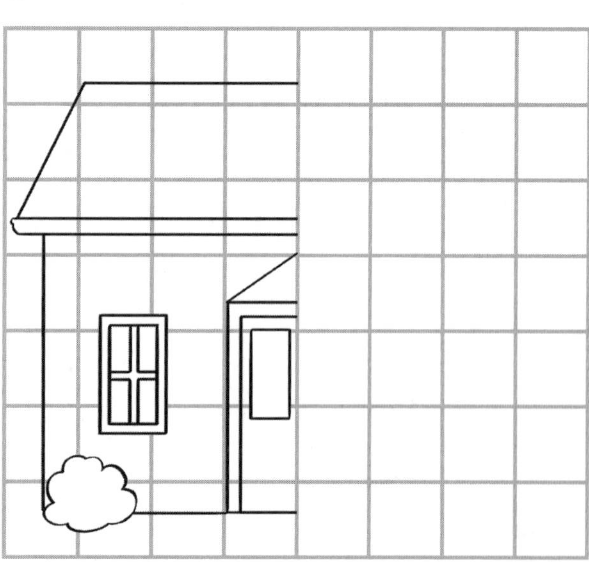

b

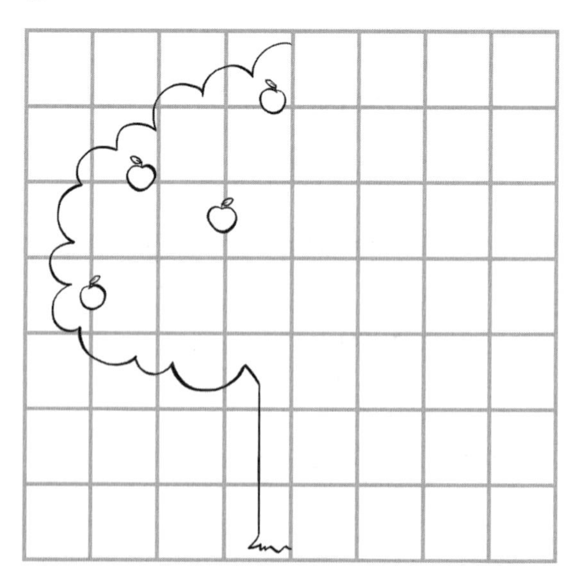

DATE:

STUDENT ASSESSMENT

1 Draw an example of each angle.

a acute **b** obtuse **c** reflex

2 Draw the lines of symmetry for each shape.

a

b

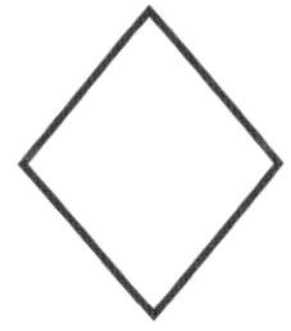

3 Find the value of each letter.

a

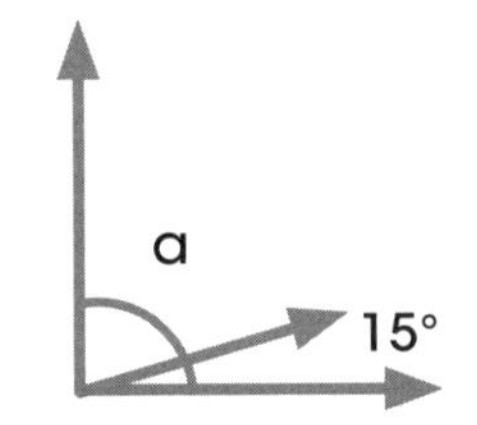

a = ______

b

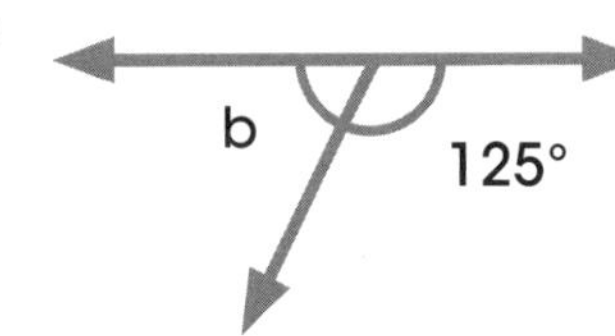

b = ______

c

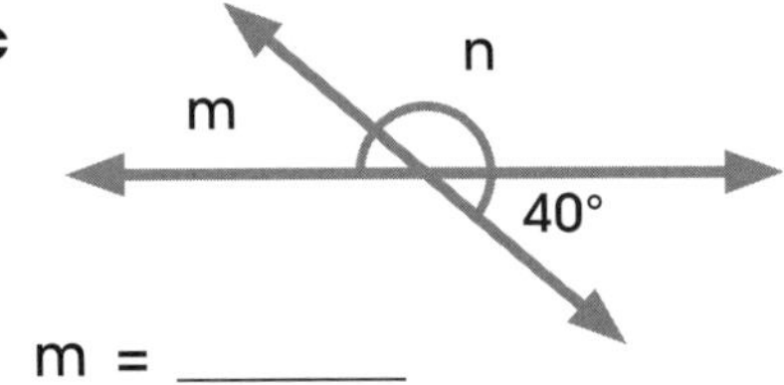

m = ______

n = ______

d

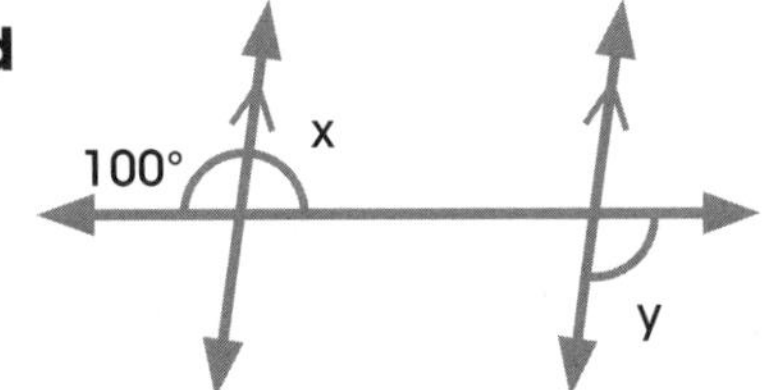

x = ______ y = ______

4 Answer the following.

a What is a right angle? ______

b What is a straight angle? ______

c How can this help you find missing angle values?

______

______

______

Challenges

Going Up Stairs

1 Draw the next 2 stair shapes on the grid.

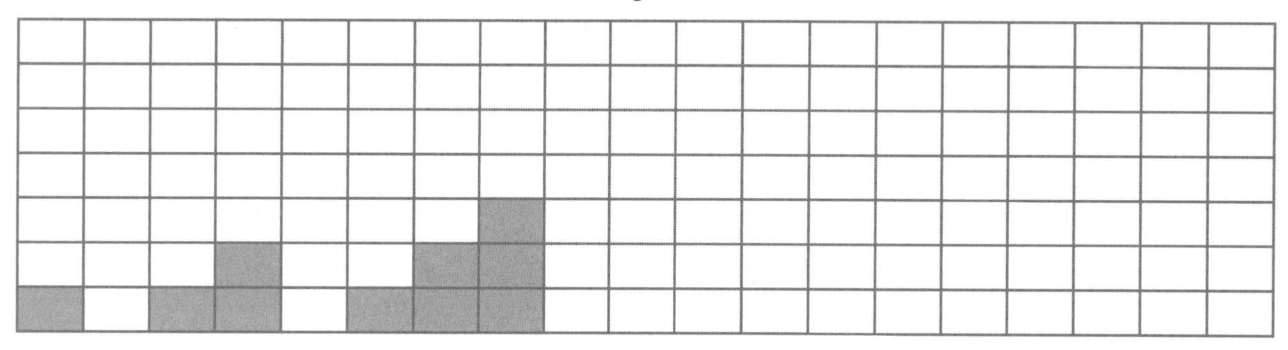

2 Complete the table to show the perimeter and area of each stair shape.

Number of stairs	1	2	3	4	5
Perimeter (units)	4				
Area (units²)	1				

Is there a pattern? ______________ Explain ____

3 Draw the tenth stair shape.

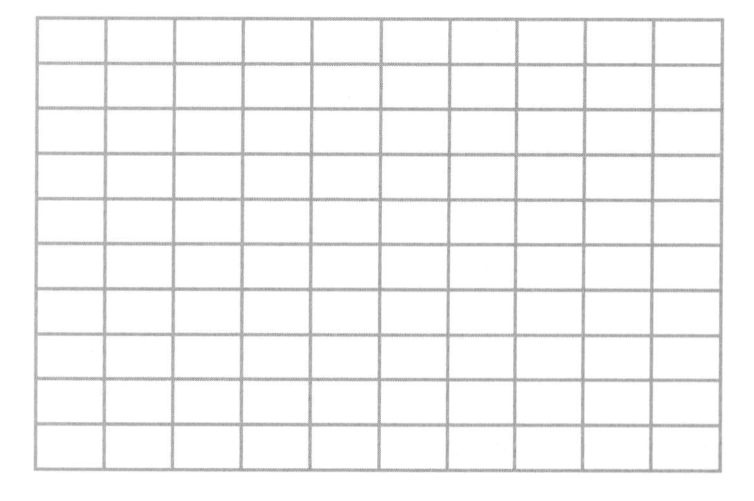

What do you notice? ____

____

Challenges

Rearranging Shapes

Change the parallelogram to a triangle, keeping the area the same.

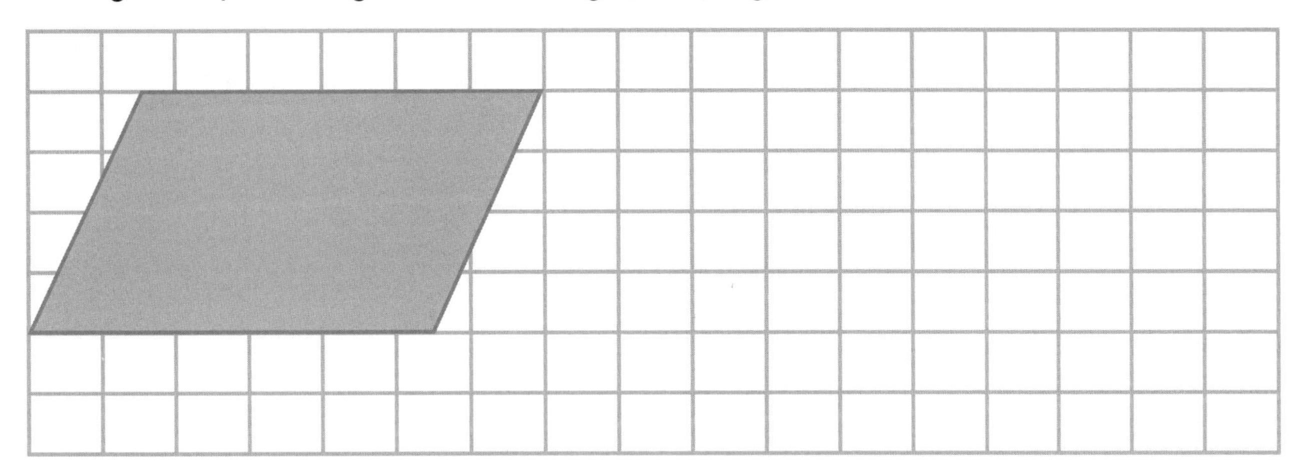

Investigating Counting

Investigate how counting began. Research the following ancient number systems:

- Mayan
- Roman
- Egyptian
- Chinese

Create an electronic display, showing your findings. Include what is similar and what is different about these systems.

Palindromic Numbers

A palindromic number is a number that is the same when its digits are reversed. For example, 11 211.

How many palindromic numbers are there between:

0 and 100? ______________________________

100 and 200? ______________________________

Challenges

Drawing Shapes

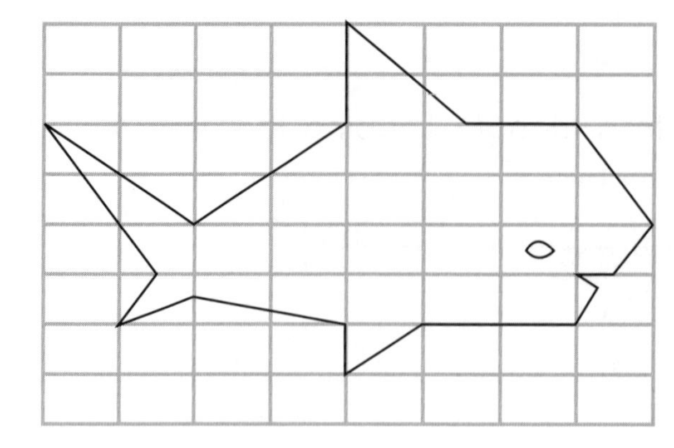

1 Use this grid to enlarge the fish.

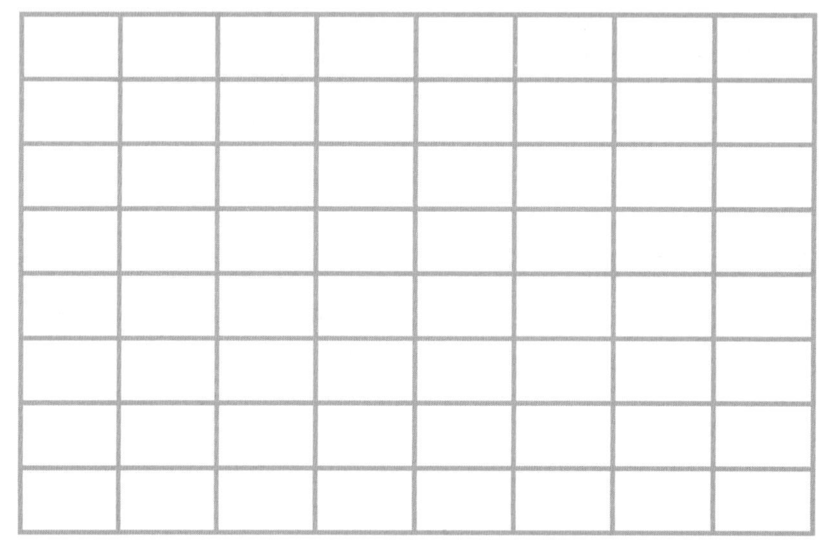

2 Use this grid to reduce the fish.

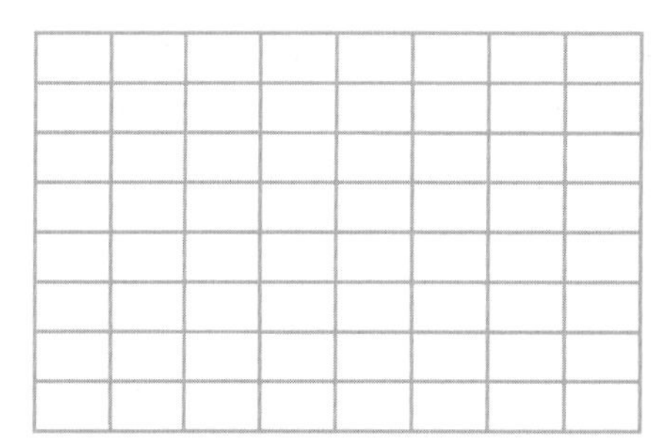

3 Try drawing the fish on an angle.

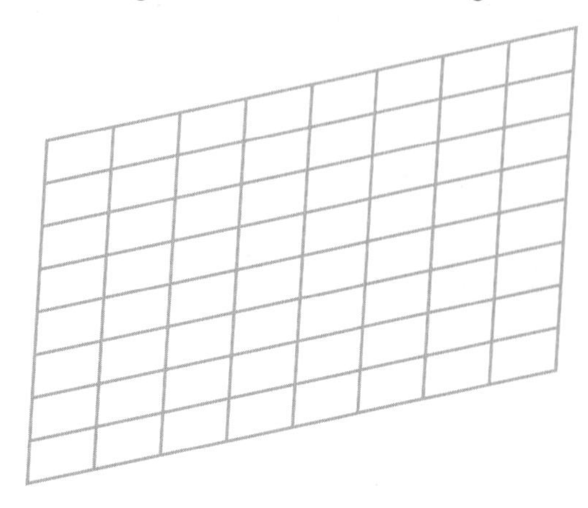

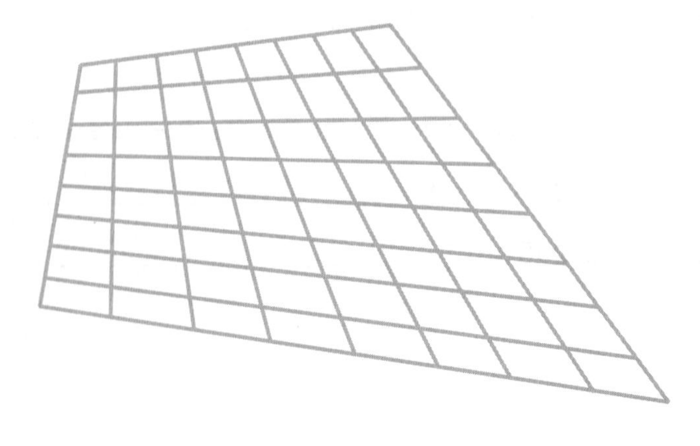

Challenges

Happy and Sad Numbers

How to find a happy number:

1 Start with any positive number.

2 Replace the number by the sum of the squares of its digits.

3 Repeat the process until the number equals 1 (this means it is a happy number). If it loops endlessly in a cycle which does not include 1, it is a sad number.

For example, for 13:

$1^2 + 3^2 = 1 + 9 = 10$

$1^2 + 0^2 = 1 + 0 = 1$

13 is happy!

Find 10 other happy numbers.

Sandwich Combinations

Alice's sandwich shop sells a basic sandwich with butter and cheese.
A customer can then select one or more of the following items to add:
lettuce, tomato, onion and egg.

Make a table with 5 columns to find out how many different kinds of sandwich Alice can make in her shop, using different combinations of ingredients.

Challenges

Triangles

How many triangles can you find in this diagram?

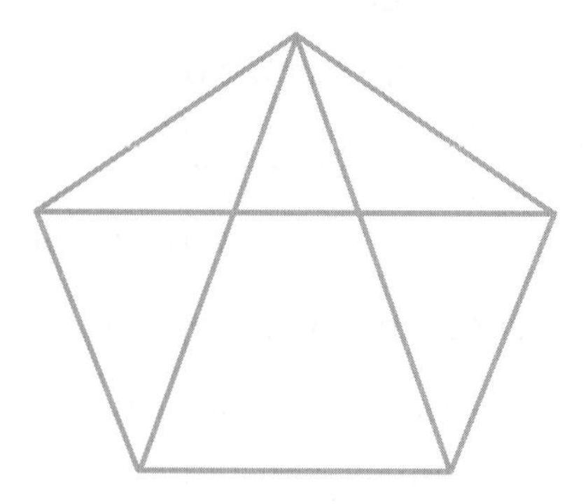

Area and Perimeter

If the area of a shape is 40 cm^2, what is the shape's perimeter? Find 10 solutions. Include a diagram for each solution.

Same Shape and Area

This grid has been divided into 2 identical parts.
Each part has the same shape and the same area.

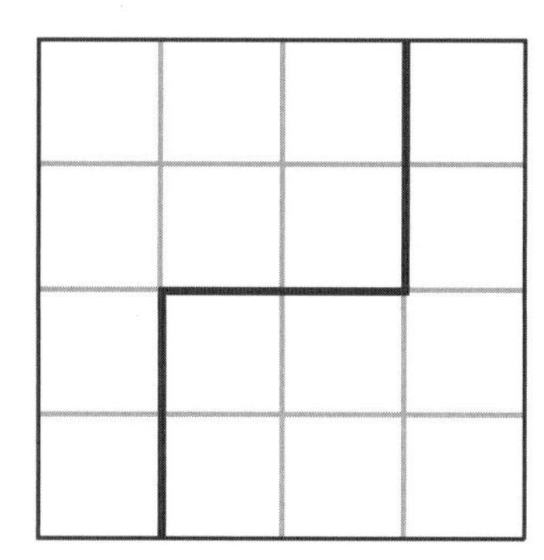

Find 4 more ways of dividing the grid into 2 identical parts. (Rotations and reflections don't count!)

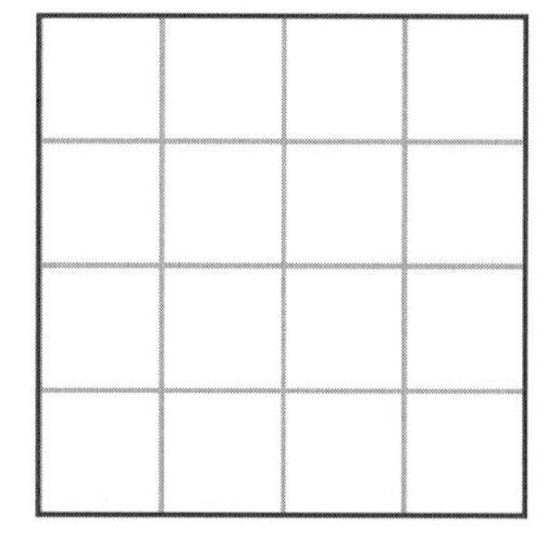

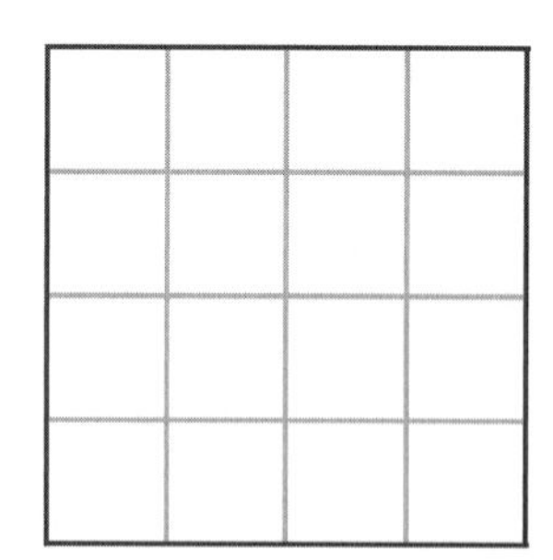

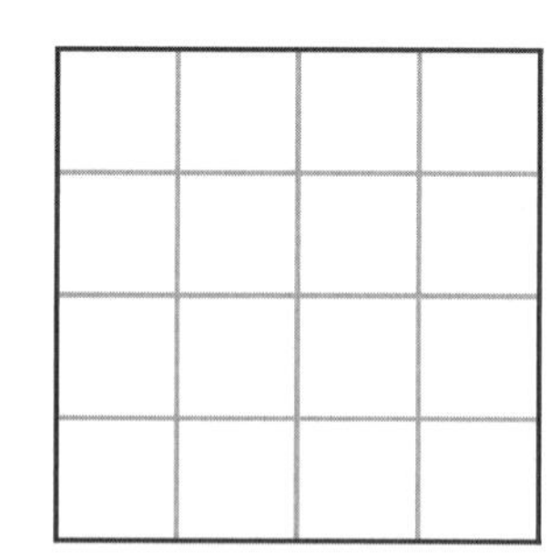

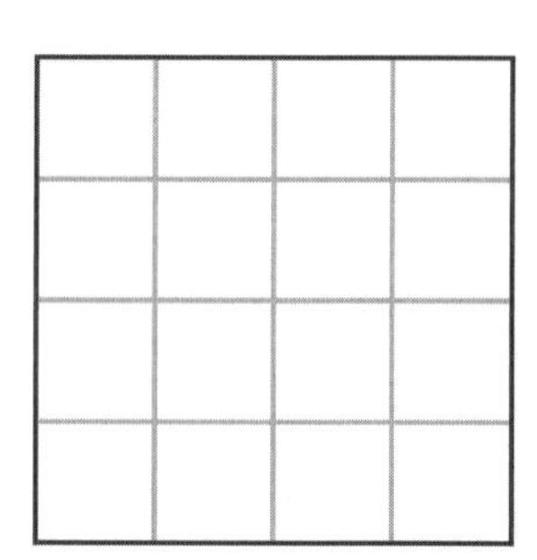

Challenges

Sudoku

Complete the sudoku so that every row, every column and every 2 × 2 square has the numbers 1, 2, 3 and 4.

	2	4	
1			3
4			2
	1	3	

Easy? Try this one. This time you need to put in the numbers 1, 2, 3, 4, 5 and 6.

		3		1	
5	6		3	2	
	5	4	2		3
2		6	4	5	
	1	2		4	5
	4		1		

Abacus

You have an abacus and 25 beads.

How many different 3-digit numbers can you make on the abacus if you must use all 25 beads in each number?

H T O

Across the River

A woman has to get a fox, a chicken, and a sack of corn across a river. She has a rowboat, but it only has room for her and one other thing. If she leaves the fox and the chicken together, the fox will eat the chicken. If she leaves the chicken and the corn together, the chicken will eat the corn. How does she get everything safely across the river?

Maths Glossary

acute angle an angle less than 90°

analogue a device that uses physical quantities rather than digits for storing and processing information, e.g. a clock with hands

angle the space between the intersection of two straight lines

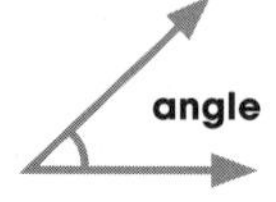

area the size of a surface; calculated with the formula: length × width

array a set of items arranged in rows and columns

axes the lines that form the framework of a graph, i.e. x-axis and y-axis

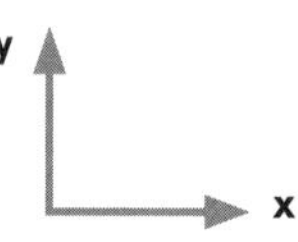

BODMAS gives the order for solving equations; it stands for: **B**rackets, **O**f, **D**ivision, **M**ultiplication, **A**ddition, **S**ubtraction

capacity how much a container can hold

chance the likelihood something will happen

circumference the distance around the outside of a circle

compass an instrument used to show direction

composite shape a shape consisting of a number of regular shapes

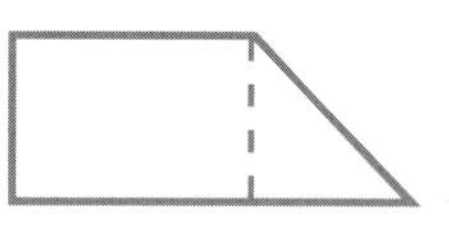

cone a 3D object with a circular base that comes to a point at the top

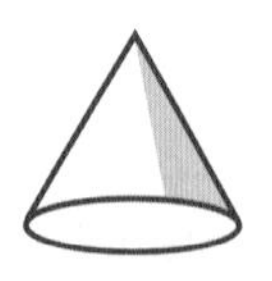

congruent 2 figures are congruent if they are exactly equal, i.e. they have the same shape and size

coordinates a pair of numbers and/or letters that represent a position on a map or graph

cross-section the shape produced when a solid shape is cut through

cube a 3D object with 6 equal square faces and 8 corners

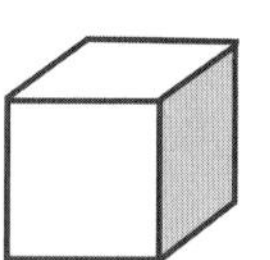

cylinder a 3D object with 2 circular faces at right angles to a curved surface

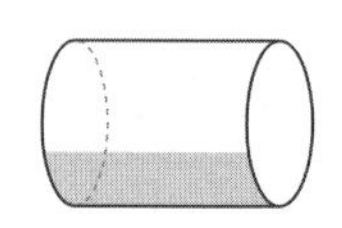

data factual information gathered for research

decimal number a number where the fraction less than zero is expressed in parts of 10, e.g. 347.15

degrees the unit for measuring an angle in geometry

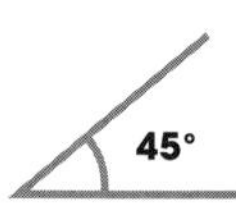

denominator the bottom number in a fraction, which shows how many parts make up a whole

difference the amount by which two numbers differ; the answer to a subtraction problem

digital describes a device that uses digits for storing and processing information, e.g. a digital clock

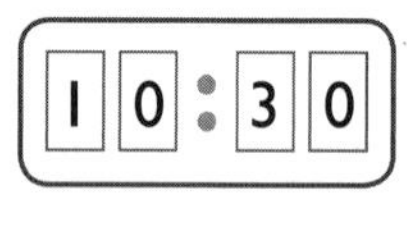

dividend a number that is divided by another number

divisor a number that divides into another number

equivalent fractions fractions that represent the same amount, e.g. $\frac{1}{4} = \frac{2}{8} = \frac{4}{16}$

Maths Glossary

factor a whole number that can be exactly divided into another number, e.g. the factors of 8 are 1, 2, 4 and 8

graph a diagram or drawing representing a collection of data. There are different types of graphs

horizontal parallel to the horizon; at right angles to the vertical direction

improper fraction a fraction with a numerator greater than the denominator

inverse operation the opposite operation, i.e. subtraction and addition or multiplication and division

irregular shape a shape that is not regular

line of symmetry a line drawn across the centre of a shape so that each half of the shape is a mirror image of the other

mass the quantity of matter in an object

mixed fraction a number made up of a whole number and a fraction, e.g. $2\frac{3}{4}$

multiple the result of multiplying a number by an integer (a number that is not a fraction), e.g. 12 is a multiple of 3

negative number a number less than zero; negative numbers are preceded by a minus sign, e.g. –4

net a flat shape that can be folded to make a 3D object

number line a line marked with numbers to show operations or patterns

number sentence a mathematical sentence written with numbers and mathematical symbols

number sequence an ordered set of numbers

numerator the top number in a fraction, which shows how many parts of the whole there are

obtuse angle an angle greater than 90° but less than 180°

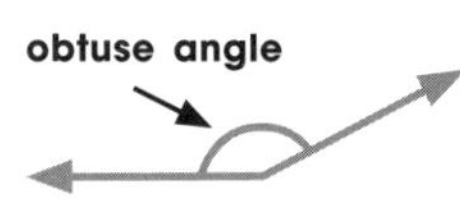

parallel lines a set of lines that remain the same distance apart and do not intersect

percentage a proportion of 100, written as %

perimeter the distance around the edge of a 2D shape

perpendicular line a vertical line that makes a right angle where it meets a horizontal line

perpendicular

place value the value of a digit dependent on its position in a number

polygon a 2D shape with 3 or more straight sides, e.g. triangle

polyhedron a 3D object with plane faces, e.g. cube, rectangular prism

positive number a number greater than zero

prime number a number that is only divisible by itself and 1, e.g. 7 is a prime number, as its only factors are 7 and 1. Note: 1 is neither a prime nor a composite number

prism a 3D object with two parallel ends of the same size and shape, e.g. rectangular prism, triangular prism

probability the chance a particular outcome will occur compared to all outcomes

product the answer to a multiplication problem

Maths Glossary

pyramid a 3D object in which the base is a polygon and all other faces are triangles

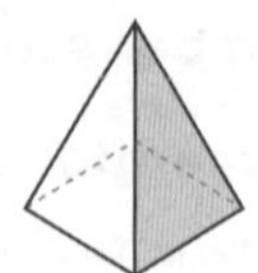

quadrilateral a plane shape that has four straight sides

quotient the answer to a division problem

reflection a mirror view of a shape or object

reflex angle an angle between 180° and 360°

regular shape a shape in which all sides are equal and all angles are equal

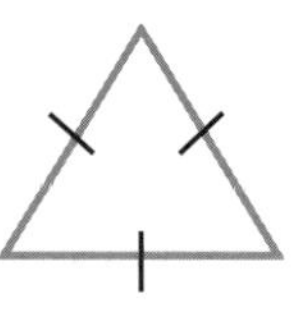

right angle an angle of exactly 90°

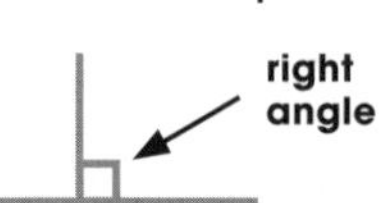

rotation turning an object around a fixed point

scale the ratio in which something is represented as either greater or smaller than lifesize

simple fraction a fraction such as $\frac{1}{2}, \frac{1}{3}, \frac{3}{7}$; also called a "common fraction" or "vulgar fraction"

sphere a 3D object shaped like a ball

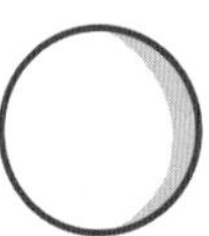

square number a number that can be represented by counters in the shape of a square, e.g. $4^2 = 4 \times 4 = 16$

sum the answer to an addition problem

symmetrical shape a shape that can be divided into 2 halves that are mirror images of each other

table information (data) organised in columns and rows

tally to keep count by making a mark to represent an item. This information may be expressed in the form of a table

three-dimensional (3D) describes an object that has length, width and height

transformation a change in position or size including: translation, rotation, reflection or enlargement (zoom)

translation the movement of a shape that occurs without flipping or reflecting it

two-dimensional (2D) describes a shape that has only 2 dimensions: length and width

unit fraction a fraction with a numerator of 1, e.g. $\frac{1}{2}$ or $\frac{1}{10}$

Venn diagram overlapping circles, used to show different sets of information

vertex the point at which 2 or more lines meet to form an angle or corner

vertical upright and at right angles to the horizontal

vertically opposite angles a pair of angles directly opposite each other formed by an intersection of straight lines

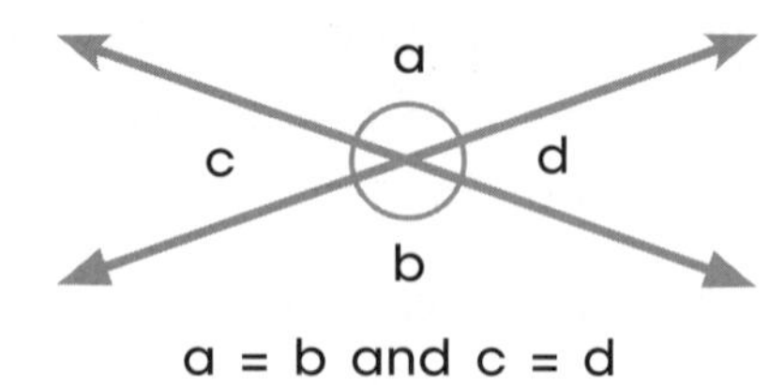

volume the space occupied by a 3D shape

weight mass affected by gravity

whole number any counting number above zero, e.g. 1, 2, 3, 4